Gnostic Anthropology

Gnostic
Anthropology

Samael Aun Weor

Gnostic Anthropology
A Glorian Book

© 2017 Glorian Publishing

Print ISBN: 978-1-943358-01-4

Originally published in Spanish as
"Antropologia Gnostica" (1978).

Glorian Publishing is a non-profit organization.
All proceeds go to further the distribution
of these books. For more information, visit
gnosticteachings.org.

Contents

Illustrations

Editor's Notes

The original edition of this book plunged the reader straight into the surging waters of the storm of debate about how mankind came to be. In this new edition, we the editors have given you some gradual steps into the pool.

Firstly, since the writing and lectures of Samael Aun Weor continually and profoundly draw from a primary resource, we provide a brief excerpt from it: *The Secret Doctrine* (1888) by H.P. Blavatsky. Being a truly monumental work, we could do no more than a short excerpt. The reader who really wants to understand what Samael Aun Weor taught is advised to study the rest of *The Secret Doctrine* as well.

Secondly, we include a lecture given by Samael Aun Weor that provides an overview of the themes of the book.

Finally, as an expansion of the points made by the author, we include an epilogue that explains the essential events of our epic history.

We hope the sum of these writings moves you towards profound reflection.

Timescale of Materialistic Science

Eon	Era	Period		Epoch	Millions of Years Ago	
Phanerozoic	Cenozoic	Quaternary		Recent	.01	Ice age ends
				Pleistocene	1.6	Ice age begins Earliest humans
		Tertiary	Neogene	Pliocene		
					5.3	
				Miocene		
					23.7	
			Paleogene	Oligocene		
					36.6	
				Eocene		
					57.8	
				Paleocene		
					66	Dinosaur extinction
	Mesozoic	Cretaceous				
					144	
		Jurassic				
					208	
		Triassic				
					245	First mammals First dinosaurs
	Paleozoic	Permian				
					286	
		Carboniferous	Pennsylvanian			
					320	First reptiles
			Mississippian			First amphibians
					360	
		Devonian				
					408	
		Silurian				
					438	First land plants
		Ordovician				First fish
					505	
		Cambrian				
					570	
Precambrian		Protezoic Eon				Earliest shelled animals
					2500	
		Archean Eon				
					3800	Earliest fossils

Dimensions and the Ray of Creation

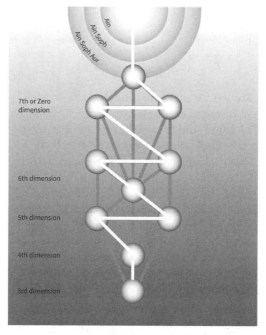

Ain

Ain Soph

Ain Soph Aur

7th or Zero
dimension

6th dimension

5th dimension

4th dimension

3rd dimension

Modern science is only capable of analyzing physical phenomena, thus its timescale [at left] merely reflects theories
about physical data. Kabbalah explains how all manifested things "condensed" from very simple, subtle forms
slowly into increased density and complexity, through a
series of "dimensions," or layers of nature. This is similar
to how we create something: (1) we have an inspiration
(which is very subtle), (2) we make a plan, (3) we start to
put the pieces together, and then finally (4) our creation
is made. We can do that because nature does that on a
larger scale. This is the basis of Gnostic anthropology.

"Modern science insists upon the doctrine of evolution; so do human reason and the "Secret Doctrine," and the idea is corroborated by the ancient legends and myths, and even by the Bible itself when it is read between the lines. We see a flower slowly developing from a bud, and the bud from its seed. But whence the latter, with all its predetermined programme of physical transformation, and its invisible, therefore spiritual forces which gradually develop its form, colour, and odour? The word evolution speaks for itself. The germ of the present human race must have preexisted in the parent of this race, as the seed, in which lies hidden the flower of next summer, was developed in the capsule of its parent flower..."
— H.P Blavatsky, "Isis Unveiled," Vol. I., p. 153

Archaic or Modern Anthropology?

Excerpted from *The Secret Doctrine* (1888)
by H. P. Blavatsky

> "The question of questions for mankind — the problem which underlies all others, and is more deeply interesting than any other — is the ascertainment of the place which man occupies in Nature, and of his relations to the Universe of things." — Huxley.

The world stands divided this day, and hesitates between *divine* progenitors — be they Adam and Eve or the lunar Pitris — and *Bathybius Haeckelii,* the gelatinous hermit of the briny deep. Having explained the occult theory, it may now be compared with that of the modern Materialism. The reader is invited to choose between the two after having judged them on their respective merits.

We may derive some consolation for the rejection of our divine ancestors, in finding that the Haeckelian speculations receive no better treatment at the hands of strictly *exact* Science than do our own.

Haeckel's *phylogenesis* is no less laughed at by the foes of his fantastic evolution, by other and greater Scientists, than our primeval races will be. As du Bois-Reymond puts it, we may believe him easily when he says that "ancestral trees of our race sketched in the 'Schopfungs-geschichte' are of about as much value as are the pedigrees of the Homeric heroes in the eyes of the historical critic."

This settled, everyone will see that one hypothesis is as good as another. And as we find that German naturalist (Haeckel) himself confessing that neither geology (in its history of the past) nor the ancestral history of organisms will ever "rise to the position of a real exact Science,"[1] a large margin is thus left to Occult Science to make its annotations and lodge its protests. The world is left to choose between the teachings of Paracelsus, the "Father of Modern Chemistry," and those of Haeckel, the Father of the mythical Sozura. We demand no more.

Without presuming to take part in the quarrel of such very learned naturalists as du Bois-Reymond and Haeckel *a propos* of

---

1   "Pedigree of Man." — "The Proofs of Evolution," p. 273.

our blood relationship to "those ancestors (of ours) which have led up from the unicellular classes, Vermes, Acrania, Pisces, Amphibia, Reptilia to the Aves" — one may put in a few words, a question or two, for the information of our readers. Availing ourselves of the opportunity, and bearing in mind Darwin's theories of natural selection, etc., we would ask Science — with regard to the origin of the human and animal species — which theory of evolution of the two herewith described is the more scientific, or the more *unscientific*, if so preferred.

1. Is it that of an Evolution which starts from the beginning with sexual propagation?

2. Or that teaching which shows the gradual development of organs; their solidification, and the procreation of each species, at first by simple easy separation from one into two or even several individuals. Then follows a fresh development — the first step to a species of separate distinct sexes — the hermaphrodite condition; then again, a kind of Parthenogenesis,

"virginal reproduction,"
when the egg-cells are formed
within the body, issuing from
it in atomic emanations and
becoming matured outside of
it; until, finally, after a definite
separation into sexes, the
human beings begin procreating
through sexual connection?

Of these two, the former "theory," —
rather, a "*revealed* fact" — is enunciated by
all the *exoteric* Bibles (except the Puranas),
preeminently by the Jewish Cosmogony.
The last one, is that which is taught by
the Occult philosophy, as explained all
along [in *The Secret Doctrine* of Blavatsky].

An answer is found to our question
in a volume just published by Mr. S.
Laing — the best lay exponent of Modern
Science. In chapter viii. of his latest work,
"A Modern Zoroastrian," the author
begins by twitting "all ancient religions
and philosophies" for "assuming a male
and female principle for their gods." At
first sight, he says "the distinction of sex
appears as fundamental as that of plant
and animal." . . . . "The Spirit of god
brooding over Chaos and producing the
world," he goes on to complain, "is only a

later edition, revised according to monotheistic ideas, of the far older Chaldean
legend which describes the creation of
Kosmos out of Chaos by the co-opera-
tions of great gods, male and female[2]... "
Thus, in the orthodox Christian creed we
are taught to repeat "begotten, not made,"
a phrase which is absolute nonsense, an
instance of using words like counterfeit
notes, which have no solid value of an
idea behind them. For "begotten" is a very
definite term which "implies the conjunction of two opposite sexes to produce a
new individual."

However we may agree with the learned
author as to the inadvisability of using
wrong words, and the terrible anthropomorphic and *phallic* element in the old
Scriptures — especially in the orthodox
Christian Bible — nevertheless, there may
be two extenuating circumstances in the
case. Firstly, all these "ancient philosophies" and "modern religions" are — as
sufficiently shown in these two volumes
[*The Secret Doctrine*] — an exoteric veil
thrown over the face of esoteric truth;

---

2   In Genesis, the word in Hebrew is Elohim, a plural
word that means "gods and goddesses," which is mis-
translated into other languages as singular masculine,
"god."

and — as the direct result of this — they are allegorical, i.e., mythological in form; but still they are immensely more philosophical in essence than any of the new *scientific* theories, so-called. Secondly, from the Orphic theogony down to Ezra's last remodelling of the Pentateuch, every old Scripture having in its origin borrowed its facts from the East, it has been subjected to constant alterations by friend and foe, until of the original version there remained but the name, a dead shell from which the Spirit had been gradually eliminated.

This alone ought to show that no religious work now extant can be understood without the help of the Archaic wisdom, the primitive foundation on which they were all built.

But to return to the direct answer expected from Science to *our* direct question. It is given by the same author, when, following his train of thought on the unscientific euhemerization of the powers of Nature in ancient creeds, he pronounces a condemnatory verdict upon them in the following terms: —

"Science, however, makes sad havoc with this impression of *sexual genera-*

*tion being the original and only mode of reproduction,** and the microscope and dissecting knife of the naturalist introduce us to new and altogether unsuspected (?) worlds of life. . . ."

So little *"unsuspected,"* indeed, that the *original a*-sexual "modes of reproduction" must have been known — to the ancient Hindus, at any rate — Mr. Laing's assertion to the contrary, notwithstanding. In view of the statement in the Vishnu Purana, quoted by us elsewhere, that Daksha "established sexual intercourse as the means of multiplication," only after a series of other "modes," which are all enumerated therein, (Vol. II., p. 12, Wilson's Transl.), it becomes difficult to deny the fact. This assertion, moreover, is found, note well, in an **exoteric** work. Then, Mr. S. Laing goes on to tell us that: —

. . . . "By far the larger proportion of living forms, in number . . . . have come into existence, *without the aid of sexual* propagation." He then instances Haeckel's monera . . . . *"multiplying by self-division."* The next stage the author shows in the nucleated cell, "which does exactly the same thing." The following stage is that in "which

the organism does not divide into
two equal parts, but a *small portion of
it swells out... and finally parts company*
and starts on separate existence,
which grows to the size of the parent
by its inherent faculty of manufacturing fresh protoplasm from surround-
ing inorganic materials."[3]

This is followed by a many-celled
organism which is formed by *"germ-buds
reduced to spores, or single cells, which are emit-
ted from the parent"*... when "we are at the
threshold of that system of sexual propagation, which has (now) become the rule
in all the higher families of animals"... It
is when an "organism, having advantages
in the struggle for life, established itself
permanently"... that special organs developed to meet the altered condition... when
a distinction "would be firmly established
of a female organ or ovary containing
the egg or primitive cell from which the
new being was to be developed."... "This is
confirmed by a study of embryology, *which
shows that in the* **human** *and higher animal
species the distinction of sex is not developed*
until a considerable progress has been

3   In this, as shown in Part I [of *The Secret Doctrine*],
Modern Science was again anticipated, far beyond its own
speculations in this direction, by *Archaic* Science.

made in the growth of the embryo..." In
the great majority of plants, and in some
lower families of animals... the male and
female organs are developed within the
same being... a hermaphrodite. Moreover,
in the "virginal reproduction — germ-cells
apparently similar in all respects to egg-
cells, develop themselves into new individuals *without any fructifying element*," etc.,
etc. (pp. 103-107).

Of all which we are as perfectly well
aware as of this — that the above was
never applied by the very learned English
popularizer of Huxleyo-Haeckelian theories to the genus *homo*. He limits this to
specks of protoplasm, plants, bees, snails,
and so on. But if he would be true to the
theory of descent, he must be as true to
ontogenesis, in which the fundamental
biogenetic law, we are told, runs as follows: "the development of the embryo
(ontogeny) is a condensed and abbrevi-
ated repetition of *the evolution of the race*
(phylogeny). This repetition is the more
complete, the more the true original
order of evolution (palingenesis) has been
retained by continual heredity. On the
other hand, this repetition is the less complete, the more by varying adaptations the

later spurious development (caenogenesis) has obtained." (Anthrop. 3rd edition, p. 11.)

This shows to us that every living creature and thing on earth, including man, evolved from *one common primal form*. Physical man must have passed through the same stages of the evolutionary process in the various modes of procreation as other animals have: he must have *divided* himself; then, hermaphrodite, have given birth *parthenogenetically* (on the immaculate principle) to his young ones; the next stage would be the *oviparous* — at first "without any fructifying element," then "with the help of the fertilitary spore"; and only after the final and definite evolution of both sexes, would he become a distinct "male and female," when reproduction through sexual union would grow into universal law. So far, all this is scientifically proven. There remains but one thing to be ascertained: the plain and comprehensively described processes of such *ante*-sexual reproduction. This is done in the Occult books, a slight outline of which was attempted by the writer in Part I. of this Volume [*The Secret Doctrine*].

Either this, or — man is a distinct being. Occult philosophy may call him that, because of his distinctly *dual* nature. Science cannot do so, once that it rejects every interference save *mechanical laws, and* admits of no principle outside matter. The former — the archaic Science — allows the human physical frame to have passed through every form, from the lowest to the very highest, its present one, or from the simple to the complex — to use the accepted terms. But it claims that in this cycle (the fourth), the frame having already existed among the types and models of nature from the preceding Rounds — that it was quite ready for man from the beginning of *this Round*.[4] The Monad had but to step into the astral body of the progenitors, in order that the work

4   Theosophists will remember that, according to occult teaching, cyclic pralayas so-called are but obscurations, during which periods Nature, i.e., everything visible and invisible on a resting planet — remains in status quo. Nature rests and slumbers, no work of destruction going on on the globe even if no active work is done. All forms, as well as their astral types, remain as they were at the last moment of its activity. The "night" of a planet has hardly any twilight preceding it. It is caught like a huge mammoth by an avalanche, and remains slumbering and frozen till the next dawn of its new day — a very short one indeed in comparison to the "Day of Brahma."

of physical consolidation should begin around the shadowy prototype.[5]

What would Science say to this? It would answer, of course, that as man appeared on earth as the latest of the mammalians, he had no need, no more than those mammals, to pass through the primitive stages of procreation as above described. His mode of procreation was already established on Earth when he appeared. In this case, we may reply: since to this day not the remotest sign of a link between man and the animal has yet been found, then (if the Occultist doctrine is to be repudiated) he must have sprung *miraculously* in nature, like a fully armed Minerva from Jupiter's brain. And in such case the Bible is right, along with other national "revelations." Hence the

5   This will be pooh-poohed, because it will not be understood by our modern men of science; but every Occultist and theosophist will easily realize the process. There can be no objective form on Earth (nor in the Universe either), without its astral prototype being first formed in Space. From Phidias down to the humblest workman in the ceramic art — a sculptor has had to create first of all a model in his mind, then sketch it in one and two dimensional lines, and then only can he reproduce it in a three dimensional or objective figure. And if the human mind is a living demonstration of such successive stages in the process of evolution — how can it be otherwise when Nature's Mind and creative powers are concerned?

scientific scorn, so freely lavished by the author of "A Modern Zoroastrian" upon ancient philosophies and *exoteric* creeds, becomes premature and uncalled for. Nor would the sudden discovery of a "missing-link"-like fossil mend matters at all. For neither one such solitary specimen nor the *scientific conclusions* thereupon, could insure its being the long-sought-for relic, i.e., that of an undeveloped, still a once *speaking man*. Something more would be required as a final proof (vide infra, Note). Besides which, even *Genesis* takes up man, her Adam of dust, only where the Secret Doctrine leaves her "Sons of God and Wisdom" and picks up the physical man of the **Third** Race. Eve is *not* "begotten," but is extracted out of Adam on the manner of "Amoeba A," contracting in the middle and splitting into Amoeba B — by division. (See p. 103, in "The Modern Zoroastrian.") Nor has human speech developed from the various animal sounds.

Haeckel's theory that "speech arose gradually from a few simple, crude animal sounds . . . ." as such "speech still remains amongst a few races of lower rank" (*Darwinian theory in "Pedigree of Man,"* p.

22) is altogether unsound, as argued by Professor Max Muller, among others. He contends that no plausible explanation has yet been given as to how the "roots" of language came into existence. A *human* brain is necessary for *human* speech. And figures relating to the size of the respective brains of man and ape show how deep is the gulf which separates the two. Vogt says that the brain of the largest ape, the gorilla, measures no more than 30.51 cubic inches; while the average brains of the flat-headed Australian natives — the lowest now in the human races — amount to 99.35 cubic inches! Figures are awkward witnesses and cannot lie. Therefore, as truly observed by Dr. F. Pfaff, whose premises are as sound and correct as his biblical conclusions are silly: — "The brain of the apes most like man, does not amount to quite a third of the brain of the lowest races *of men: it is not half the size of the brain of a new-born child.*" (*"The Age and Origin of Man."*) From the foregoing it is thus very easy to perceive that in order to prove the Huxley-Haeckelian theories of the descent of man, it is not *one*, but a great number of *"missing links"* — a true ladder of progressive evolutionary steps

— that would have to be first found and then presented by Science to thinking and reasoning humanity, before it would abandon belief in gods and the immortal Soul for the worship of Quadrumanic ancestors. Mere myths are now greeted as "axiomatic truths." Even Alfred Russel Wallace maintains with Haeckel that primitive man was a speechless ape-creature. To this Joly answers: — "Man never was, in my opinion, this *pithecanthropus alalus* whose portrait Haeckel has drawn *as if he had seen and known him,* whose *singular* and *completely hypothetical genealogy* he has even given, from the mere mass of living protoplasm to the man endowed with speech and a civilization analogous to that of the Australians and Papuans." ("Man before Metals," p. 320, N. Joly. Inter. Scient. Series.)

Haeckel, among other things, often comes into direct conflict with the Science of languages. In the course of his attack on Evolutionism (1873, "Mr. Darwin's Philosophy of Language"), Prof. Max Muller stigmatized the Darwinian theory as "vulnerable at the beginning and at the end." The fact is, that only the partial truth of many of the *secondary*

"laws" of Darwinism is beyond question
— M. de Quatrefages evidently accepting
"Natural Selection," the "struggle for existence" and transformation within species,
as proven not once and for ever, but pro.
tem. But it may not be amiss, perhaps, to
condense the linguistic case against the
"Ape ancestor" theory: —

Languages have their phases of growth,
etc., like all else in nature. It is almost certain that the great linguistic families pass
through three stages.

1. All words are roots and
   merely placed in juxtaposition
   (Radical languages).

2. One root defines the other, and
   becomes merely a determinative
   element (Agglutinative).

3. The determinative element
   (the determinating meaning
   of which has long lapsed)
   unites into a whole with the
   formative element (Inflected).

The problem then is: Whence these
roots? Max Muller argues that the existence of these *ready-made materials of speech*
is a proof that man cannot be the crown
of a long organic series. This *potentiality of*

forming roots is the great crux which materialists almost invariably avoid.

Von Hartmann explains it as a manifestation of the "Unconscious," and admits its cogency *versus* mechanical Atheism. Hartmann is a fair representative of the Metaphysician and Idealist of the present age.

The argument has never been met by the non-pantheistic Evolutionists. To say with Schmidt: "Forsooth are we to halt before the origin of language?" is an avowal of dogmatism and of speedy defeat. (Cf. his *"Doctrine of Descent and Darwinism,"* p. 304.)

We respect those men of science who, wise in their generation, say: "Prehistoric Past being utterly beyond our powers of direct observation, we are too honest, too devoted to the truth — or what we regard as truth — to speculate upon the unknown, giving out our unproven theories along with facts absolutely established in modern Science." . . . . "The borderland of (metaphysical) knowledge is best left to time, which is the best test as to truth" (*A Modern Zoroastrian*, p. 136).

This is a wise and an honest sentence in the mouth of a materialist. But when

a Haeckel, after just saying that "*historical*
events of past time . . " having "occurred
many *millions of years ago*,[6] . . . are for ever
removed from direct observation," and
that neither geology nor phylogeny[7] can
or will "rise to the position of a real exact
science," then insists on the development
of all organisms — "from the lowest vertebrate to the highest, from Amphioxus
to man" — we ask for a weightier proof
than he can give. Mere "*empirical* sources
of knowledge," so extolled by the author
of "*Anthropogeny*" — when he has to be satisfied with the qualification for his own
views — are not competent to settle problems lying beyond their domain; nor is it
the province of exact science to place any
reliance on them.[8] If "empirical" — and

6   It thus appears that in its anxiety to prove our noble
descent from the catarrhine "baboon," Haeckel's school
has pushed the times of pre-historic man millions of years
back. (See "Pedigree of Man," p. 273.) Occultists, render
thanks to science for such corroboration of our claims!

7   This seems a poor compliment to pay Geology,
which is not a speculative but as exact a science as
astronomy — save, perhaps its too risky chronological
speculations. It is mainly a "Descriptive" as opposed to an
"Abstract" Science.

8   Such newly-coined words as "perigenesis of plas-
tids," "plastidule Souls" (!), and others less comely,
invented by Haeckel, may be very learned and correct in
so far as they may express very graphically the ideas in his
own vivid fancy. As a fact, however, they remain for his

Haeckel declares so himself repeatedly — then they are no better, nor any more reliable, in the sight of exact research, when extended into the remote past, than our Occult teachings of the East, both having to be placed on quite the same level. Nor are his *phylogenetic* and *palingenetic* speculations treated in any better way by the real scientists, than are our cyclic repetitions of the evolution of the Great in the minor races, and the original order of evolutions. For the province of exact, real Science, materialistic though it be, is to carefully avoid anything like guess-work, speculation which *cannot* be verified; in short, all *suppressio veri* and all *suggestio falsi*. The business of the man of exact Science is to observe, each in his chosen department, the phenomena of nature; to record, tabu-

---

less imaginative colleagues painfully caenogenetic — to use his own terminology; i.e., for true Science they are spurious speculations so long as they are derived from "empirical sources." Therefore, when he seeks to prove that "the origin of man from other mammals, and most directly from the catarrhine ape, is a deductive law that follows necessarily from the inductive law of the theory of descent" ("Anthropogeny," p. 392) — his no less learned foes (du Bois-Reymond — for one) have a right to see in this sentence a mere jugglery of words; a "testimonium paupertatis of natural science" — as he himself complains, calling them, in return, ignoramuses (see "Pedigree of Man," Notes).

late, compare and classify the facts, down to the smallest minutiae *which are presented to the observation of the senses with the help of all the exquisite mechanism that modern invention supplies, not by the aid of metaphysical flights of fancy.* All that he has a legitimate right to do, is to correct by the assistance of physical instruments the defects or illusions of his own coarser vision, auditory powers, and other senses. He has no right to trespass on the grounds of metaphysics and psychology. His duty is to verify and to rectify all the facts that *fall under his direct* observation; to profit by the experiences and mistakes of the Past in endeavouring to trace the working of a certain concatenation of cause and effects, which, but only by its constant and unvarying repetition, may be called **a Law**. This it is which a man of science is expected to do, if he would become a teacher of men and remain true to his original programme of natural or physical sciences. Any side-way path from this royal road becomes *speculation*.

Instead of keeping to this, what does many a so-called man of science do in these days? He rushes into the domains of pure metaphysics, while deriding it.

He delights in rash conclusions and calls
it "a *deductive* law from the *inductive* law"
of a theory based upon and drawn out of
the depths of his own consciousness: that
consciousness being perverted by, and
honeycombed with, one-sided materialism. He attempts to explain the "origin"
of things, which are yet embosomed only
in his own conceptions. He attacks spiritual beliefs and religious traditions mil-
lenniums old, and denounces everything,
save his own hobbies, as superstition.
He suggests theories of the Universe, a
Cosmogony developed by blind, mechanical forces of nature alone, far more *miraculous and impossible* than even one based
upon the assumption of *fiat lux* out of
nihil — and tries to astonish the world by
such a wild theory; which, being known to
emanate from a scientific brain, is taken
on blind faith as very scientific and the
outcome of **Science**.

Are those the opponents Occultism
would dread? Most decidedly not. For
such theories are no better treated by *real*
(not empirical) Science than our own.
Haeckel, hurt in his vanity by du Bois-
Reymond, never tires of complaining
publicly of the latter's onslaught on his

fantastic theory of descent. Rhapsodizing on "the exceedingly rich storehouse of empirical evidence," he calls those "recognised physiologists" who oppose every speculation of his drawn from the said "storehouse" — *ignorant* men. "If many men," he declares — "and among them even some scientists of repute — hold that the whole of phylogeny is a castle in the air, and genealogical trees (from monkeys?) are empty plays of phantasy, they only in speaking thus demonstrate their ignorance of that wealth of *empirical sources of knowledge* to which reference has already been made" ("Pedigree of Man," p. 273).

We open Webster's Dictionary and read the definitions of the word "empirical": "Depending upon experience or observation alone, *without due regard to modern science and theory.*" This applies to the Occultists, Spiritualists, Mystics, etc., etc. Again, "an Empiric — One who confines himself to applying the results of his own observations" (only) (which is Haeckel's case); "one *wanting [lacking] Science* . . . . an ignorant and unlicensed practitioner; a quack; a **Charlatan**."

No Occultist or "magician," has ever been treated to any worse epithets. Yet the Occultist remains on his own metaphysical grounds, and does not endeavour to rank *his knowledge*, the fruits of *his* personal observation and experience, among the *exact* sciences of modern learning. He keeps within his legitimate sphere, where he is master. But what is one to think of a rank materialist, whose duty is clearly traced before him, who uses such an expression as this: —

"The origin of man from other mammals, and most directly from the catarrhine ape, *is a deductive law, that follows necessarily from the inductive law of the* **Theory of Descent**." ("Anthropogeny," p. 392).

A "theory" is simply a hypothesis, a speculation, *and no law*. To say otherwise is only one of the many liberties taken now-a-days by scientists. They enunciate an absurdity, and then hide it behind the shield of Science. Any deduction from theoretical speculation is no better than *a speculation on a speculation*. Now Sir W. Hamilton has already shown that the word theory is now used "in a very loose and improper sense"... "that it is convertible into *hypothesis*, and *hypothesis* is commonly used as another term for *conjecture*, whereas the terms 'theory' and 'theoretical' are properly used in opposition to the term *practice* and *practical*."

But modern Science puts an extinguisher on the latter statement, and mocks at the idea. Materialistic philosophers and Idealists of Europe and America may be agreed with the Evolutionists as to the physical origin of man — yet it will never become a general truth with the true metaphysician, and the latter defies the materialists to make good their arbitrary assumptions. That the ape-theory theme[9] of Vogt and Darwin, on which the Huxley-Haeckelians have composed of

9   The mental barrier between man and ape, characterized by Huxley as an "enormous gap, a distance practically immeasurable"! ! is, indeed, in itself conclusive. Certainly it constitutes a standing puzzle to the materialist, who relies on the frail reed of "natural selection." The physiological differences between Man and the Apes are in reality — despite a curious community of certain features— equally striking. Says Dr. Schweinfurth, one of the most cautious and experienced of naturalists: —
"In modern times there are no animals in creation that have attracted more attention from the scientific student than the great quadrumana (the anthropoids), bearing such a striking resemblance to the human form as to have justified the epithet of anthropomorphic being conferred on them. . . . But all investigation at present only leads human intelligence to a confession of its insufficiency; and nowhere is caution more to be advocated, nowhere is premature judgment more to be deprecated than in the

late such extraordinary variations, is far less scientific — because clashing with the fundamental laws of that theme itself — than ours can ever be shown to be, is very easy of demonstration. Let the reader only turn to the excellent work on "Human Species" by the great French naturalist de Quatrefages, and our statement will at once be verified.

Moreover, between the esoteric teaching concerning the origin of man and Darwin's speculations, no man, unless he is a rank materialist, will hesitate. This is the description given by Mr. Darwin of "the earliest ancestors of man."

> "They were without doubt once covered with hair; both sexes having beards; their ears were pointed and capable of movement; and their bodies were provided with a tail, having the proper muscles. Their limbs and bodies were acted on by many muscles which now only occasionally reappear in man, but which are still normally present in the quadrumana. . . . The foot, judging from the condition of the great toe in the foetus, was

---

attempt to bridge over the mysterious chasm which separates man and beast." "Heart of Africa" i., 520.

then prehensile, and our progenitors, no doubt, were arboreal in their habits, frequenting some warm forest-clad land, and the males were provided with canine teeth which served as formidable weapons. . . ."[10]

Darwin connects him with the type of the tailed catarrhines, "and consequently removes him a stage backward in the scale of evolution. The English naturalist is not satisfied to take his stand upon the ground of his own doctrines, and, like Haeckel, on this point places himself in direct variance with one of the fundamental laws which constitute the principal charm of Darwinism . . . " And then the learned French naturalist proceeds to show how this fundamental law is broken. "In fact," he says, "in the theory of Darwin, transmutations do not take place, either by chance or in every direction. They are ruled by certain laws which are due to the organization itself. If an organism is once modified in a

10   A ridiculous instance of evolutionist contradictions is afforded by Schmidt ("Doctrine of Descent and Darwinism," on page 292). He says, "Man's kinship with the apes is not impugned by the bestial strength of the teeth of the male orang or gorilla." Mr. Darwin, on the contrary, endows this fabulous being with teeth used as weapons!

given direction, it can undergo secondary or tertiary transmutations, but will still preserve the impress of the original. It is the law of permanent characterization, which alone permits Darwin to explain the filiation of groups, their characteristics, and their numerous relations. It is by virtue of this law that all the descendants of the first mollusc have been molluscs; all the descendants of the first vertebrate have been vertebrates. It is clear that this constitutes one of the foundations of the doctrine. . . . It follows that two beings belonging to two distinct types can be referred to a common ancestor, but the one cannot be the descendant of the other"; (p. 106).

"Now man and ape present a very striking contrast in respect to type. Their organs . . . correspond almost exactly term for term: but these organs are arranged after a very different plan. In man they are so arranged that he is essentially a walker, while in apes they necessitate his being a climber. . . . There is here an anatomical and mechanical distinction. . . . A glance at the page where Huxley has figured side by side a human skeleton and the skeletons of the most highly developed apes is a sufficiently convincing proof."

The consequence of these facts, from the point of view of the logical application of the law of permanent characterizations, is that man cannot be descended from an ancestor who is already characterized as an ape, any more than a catarrhine tailless ape can be descended from a tailed catarrhine. A walking animal cannot be descended from a climbing one.

"Vogt, in placing man among the primates, declares without hesitation that the lowest class of apes have passed the landmark (the common ancestor), from which the different types of this family have originated and diverged." (This ancestor of the apes, occult science sees in the lowest human group during the Atlantean period, as shown before.) ... "We must, then, place the origin of man beyond the last apes," goes on de Quatrefages, thus corroborating our Doctrine, "if we would adhere to one of the laws most emphatically necessary to the Darwinian theory. We then come to the prosimiae of

Haeckel, the loris, indris, etc. But those animals also are climbers; we must go further, therefore, in search of our first direct ancestor. But the genealogy by Haeckel brings us from the latter to the marsupials. . . . From men to the Kangaroo the distance is certainly great. Now neither living nor extinct fauna show the intermediate types which ought to serve as landmarks. This difficulty causes but slight embarrassment to Darwin.[11] We know that he considers the want of information upon similar questions as a proof in his favour. Haeckel doubtless is as little embarrassed. He admits the existence of an absolutely theoretical pithecoid man."

"Thus, since it has been proved that, according to Darwinism itself, the origin of man must be placed beyond the eighteenth stage, and since it becomes, in consequence, necessary to fill up the gap between marsupials and man, will Haeckel admit the

11   According even to a fellow-thinker, Professor Schmidt, Darwin has evolved "a certainly not flattering, and perhaps in many points an incorrect, portrait of our presumptive ancestors in the dawn of humanity." ("Doctrine of Descent and Darwinism," p. 284.)

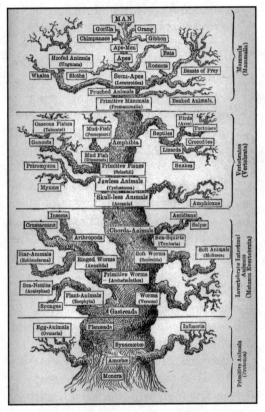

"The Pedigree of Man" from Haeckel's "Evolution of Man"

existence of four unknown intermediate groups instead of one?" asks de Quatrefages. "Will he complete his genealogy in this manner? It is not for me to answer." ("The Human Species," p. 107-108.)

But see Haeckel's famous genealogy, in "The Pedigree of Man," called by him "Ancestral Series of Man." In the "Second Division" (Eighteenth Stage) he describes "Prosimiae, allied to the Loris (Stenops) and Makis (Lemur) as without marsupial bones and cloaca, but with placenta." And now turn to de Quatrefages' "The Human Species," pp. 109, 110, and see his proofs, based on the latest discoveries, to show that "the prosimiae of Haeckel have no decidua and a diffuse placenta." They cannot be the ancestors of the apes even, let alone man, according to a fundamental law of Darwin himself, as the great French Naturalist shows. But this does not dismay the "animal theorists" in the least, for self-contradiction and paradoxes are the very soul of modern Darwinism. Witness — Mr. Huxley. Having himself shown, with regard to fossil man and the "missing link," that "neither in quaternary ages nor at the present time does any intermediary being fill the gap which separates man from the Troglodyte"; and that to "deny the existence of this gap would be as reprehensible as absurd," the great man of Science denies his own words in actu by supporting with all the weight of his scientific authority that most "absurd" of all theories — the descent of man from an ape!

"This genealogy," says de Quatrefages, "is wrong throughout, and is founded on a material error." Indeed, Haeckel bases his descent of man on the 17th and 18th stages (See Aveling's "Pedigree of Man," p. 77), the marsupialia and prosimiae — (genus Haeckelii?). Applying the latter term to the Lemuridae — hence making of them animals with a placenta — he commits a zoological blunder. For after having himself divided mammals according to their anatomical differences into two groups: the indeciduata, which have no decidua (or special membrane uniting the placentae), and the deciduata, those who possess it: he includes the prosimiae in the latter group. Now we have shown elsewhere what other men of science had to say to this. As de Quatrefages says, "The anatomical investigations of . . . Milne

Edwards and Grandidier upon these animals . . . place it beyond all doubt that the prosimiae of Haeckel have no decidua and a diffuse placenta. They are indeciduata. Far from any possibility of their being the ancestors of the apes, according to the principles laid down by Haeckel himself, they cannot be regarded even as the ancestors of the zonoplacental mammals . . . and ought to be connected with the pachydermata, the edentata, and the cetacea"; (p. 110). And yet Haeckel's inventions pass off with some as exact science!

The above mistake, if indeed, one, is not even hinted at in Haeckel's "Pedigree of Man," translated by Aveling. If the excuse may stand good that at the time the famous "genealogies" were made, "the embryogenesis of the prosimiae was not known," it is familiar now. We shall see whether the next edition of Aveling's translation will have this important error rectified, or if the 17th and 18th stages remain as they are to blind the profane, as one of the real intermediate links. But, as the French naturalist observes — "their (Darwin's and Haeckel's) process is always the same, considering the unknown as a proof in favour of their theory." (Ibid.)

It comes to this. Grant to man an immortal Spirit and Soul; endow the whole animate and inanimate creation with the monadic principle gradually evolving from the latent and passive into active and positive polarity — and Haeckel will not have a leg to stand upon, whatever his admirers may say.

But there are important divergences even between Darwin and Haeckel. While the former makes us proceed from the tailed catarrhine, Haeckel traces our hypothetical ancestor to the tailless ape, though, at the same time, he places him in a hypothetical "stage" immediately preceding this: "Menocerca with tails" (19th stage).

Nevertheless, we have one thing in common with the Darwinian school: it is the law of gradual and extremely slow evolution, embracing many million years. The chief quarrel, it appears, is with regard to the nature of the primitive "Ancestor." We shall be told that the Dhyan Chohan, or the "progenitor" of Manu, is a hypothetical being unknown on the physical plane. We reply that it was believed in by the whole of antiquity, and by nine-tenths of the present humanity; whereas not only is the pithecoid man, or "ape-man," a purely hypothetical creature of Haeckel's creation, unknown and untraceable on this earth, but further its genealogy — as invented by him— clashes with scientific facts and all the known data of modern discovery in Zoology. It is simply absurd, even as a fiction. As de Quatrefages demonstrates in a few words, Haeckel "admits the existence of an absolutely theoretical pithecoid man" — a hundred times more difficult to accept than any Deva ancestor. And it is not the only instance in which he proceeds in a similar manner in order to complete his genealogical table; and he admits very naively his inventions himself. Does he not confess the non-existence of his sozura (14th stage) — a creature entirely unknown to science — by confessing over his own signature, that — "The proof of its existence arises from the necessity of an intermediate type between the 13th and the 14th stages"!

If so, we might maintain with as much scientific right, that the proof of the existence of our three ethereal races, and the three-eyed men of the Third and Fourth Root-Races "arises also from the necessity of an intermediate type" between the animal and the gods. What reason would the Haeckelians have to protest in this special case?

Of course there is a ready answer: "Because we do not grant the presence of the monadic essence." The manifestation of the Logos[12] as individual consciousness in the animal and human creation is not accepted by exact science, nor does it cover the whole ground, of course. But the failures of science and its arbitrary assumptions are far greater on the whole than any "extravagant" esoteric doctrine can ever furnish.

---

12   Greek, "word," a reference to divinity, as in "In the beginning was the Logos, and the Logos was with Theos, and the Logos was Theos." —John 1:1

Prologue

Evolution and Devolution

Presently, many philosophical doctrines based on the dogma of evolution are being taught in the eastern as well as in the western parts of the world.

Evolution and devolution are mechanical forces that are simultaneously processed in all of nature. Gnostics do not deny the reality of these two forces; they explain them.

Therefore, I, Samael Aun Weor, want to give precise data about the enigma of the human being, something that is necessary to know in order to have a better knowledge about the origin of the human being.

It is necessary to know the origin of the human being: where the human being comes from, the fundamental reason for its existence. It is necessary to investigate it deeply, in a more profound way, since only vague theories have been stated about the origin of the human being.

The present materialistic scientific bases of evolution are:

First, the nebular theories related with the origin of the universe with all of their innumerable alterations, modifications, additions, restrictions, and so on. Indeed, all of these do not even minutely alter this original mistaken conceptual theory related with the mechanical constructive processes.

Second, the capricious theory of Darwin related to the origin of the species with all its corrections and subsequent changes.

Indeed, the appearance of new species (as a result of the law of evolution) is merely a simple hypothesis, since it has never been verified. No one has ever witnessed the appearance of a new species.

When creating the theory of evolution, modern thought forgot about the destructive, devolving processes of nature. The reason lies in the much too limited field of intellectual vision of these times. Due to this, pompous theories are elaborated upon; all of them are very beautiful but all have an insufficient number of facts. Within these theories none of the processes are certainly and entirely known; these theories only observe part of the process of evolution. Erudites state that this process consists as a type of evolving change.

In this day and age, the human mind is already so degenerated that it has become incapable of even comprehending the inverse, degenerative process on a greater scale.

The minds of contemporary erudites are so trapped, bottled up within the dogma of evolution, that their minds only know how to think or function according to their bottled up condition. This is why they attribute unto the other phenomena (that is, devolution, destruction, deca- dence, and degeneration) the qualities of evolution, development, and progress.

The so-called European primitive races of the Stone Age (such as the Cro-magnon that lived in the caverns of the Earth) were formerly very beautiful, but the degener- ative, descending cyclical impulse acted terribly upon those races of Atlantean origin. Finally, the Palaeolithic man left his position to his successor, disappearing almost entirely from the scene.

None of the truly primitive villages found by explorers have shown any sign of evolution; on the contrary, in all of the cases without exception the unmistakable signs of degeneration and devolution have been observed.

Alongside any evolving process there is a degenerative, devolving process. The law of evolution and its twin sister the law of devolution work in a synchronized and harmonious way within all of creation.

From the rigorously academic point of view the word evolution means development, construction, progression, improvement, advancement, edification, etc. By focusing on the purely orthodox grammatical term, we can clearly state that devolution means reverse progression, backward motion, destruction, degeneration, decadence, etc.

Obviously, the law of antitheses coexists within any natural process. It is urgent to emphasize this transcendental idea, because the contents of this concept related with the law of antitheses are absolutely unimpeachable, unanswerable, and irrefutable. For example: day and night, light and darkness, construction and destruction, growth and fall, birth and death, etc.

Therefore, the exclusion of either of those two aforementioned laws (evolution and devolution) would produce a radical paralysis of the natural mechanism; thus to refute any of those two ordinances means, in fact, to fall into barbarism.

There is evolution in the plant that germinates, develops, and grows. There is devolution in the plant that slowly grows old and falls, becoming a pile of dust.

There is evolution in all organisms that gestate, are born, and develop. There is devolution in all creatures that become senile and die.

There is evolution in any cosmic unity that arises from the chaos. There is devolution in every planet in a consummate state, destined to become a moon, a dead body. There is evolution within all ascending civilizations, and there is devolution in every descending culture.

It is obvious that these two aforementioned laws constitute the fundamental mechanical axis of nature. Unquestionably, without such basic laws the axis of the wheel of natural mechanisms would not be able to rotate. *Life is processed in surges that rotate.*

Many people believe that small monkeys, apes, orangutans, and gorillas are evolving creatures. Some of them suppose that the human being came from the ape, but this concept fails when we observe the customs of these species of animals. Put an ape inside of a laboratory and observe what happens.

The diverse families of apes are *devolutions* that *descend* from the intellectual humanoid. The humanoid does not come from the ape, the truth of this is the inverse: *apes are degenerated humanoids.*

On earth there lives a population of about four and a half billion people; obviously these people belong to the Aryan root race.[1] Our continents are densely populated: Europe, America, Asia, Africa, Oceania (Australia and the Pacific Islands) are the five continents upon which this Aryan humanity lives.

---

1   The term "Aryan" refers to contemporary humanity. "The Aryan races, for instance, now varying from dark brown, almost black, red-brown-yellow, down to the whitest creamy colour, are yet all of one and the same stock — the Fifth Root-Race — and spring from one single progenitor, (...) who is said to have lived over 18,000,000 years ago, and also 850,000 years ago — at the time of the sinking of the last remnants of the great continent of Atlantis." —Blavatsky, The Secret Doctrine, the Synthesis of Science, Religion and Philosophy, Vol. II, p.249

If we ask, what is the origin of this Aryan humanity? Where did it come from? Do you think this Aryan humanity (which lives on these five continents) was originated on these five continents?

Human traces were found in the caves of Grimaldi. There has been an effort to reconstruct the history, or the pre-history, of the races of Grimaldi and Cro-magnon.

Skeletons of giants have been found in different places. In Brazil a figure or a human skeleton of about six or seven meters in height was found. Also skeletons of human beings that look just like gorillas, orangutans, or something of the sort, have been found, especially within the caves of Cro-magnon.

Based upon the above mentioned traces, ignoramuses have mistakenly deduced that the human race could have possibly originated from apes or monkeys.

Darwin's theory was very celebrated in his era. It was through that theory that ignoramuses started to state that the human being came from apes. This (human-ape) mystery worries our humanity. Every so often someone tries to figure out if the human being came from the ape or if the ape came from the human being. Who came from whom? During some epochs this controversy calmed down; in others it re-emerged again.

One pseudo-scientist (a rather "spoiled child" type) had the idea that the human race descended from savages; this is what he thought, and of course, in the end, his theory did not resolve anything. Who came from whom?

I do not think that the current world population of four and a half billion people originated upon these five continents, because the world has changed its geographical aspects many times. Before the Earth had the geographical aspect that you currently observe on a map, its physiognomy was different. There are older maps; there are different maps that have been found in various parts of the world where the physiognomy of the planet Earth appears to be different. The Earth did not always have the same continents, the same physiognomy. In older times, it had a different aspect. What today are the poles used to be the equator, and what today is the equator at one time was the poles.

On those maps, the present continents did not exist, or they existed only partially (they emerged after, from within the oceans). A densely populated continent that was located in the Atlantic Ocean is shown on those maps.

Therefore, the physiognomy of the planet was different. I cannot for a minute believe that the human race originated on these present continents.

The human race that developed in ancient Atlantis was very different. Remnants of the Atlantean root race were found in the caves of Cro-magnon and Grimaldi (and other caves). The apes or apemen that were found were the degenerated descendants, the degenerated remnants, of the Atlantean root race.

I state that just as there is evolution and devolution in plants and animals, and just as there is evolution and devolution in human beings, etc., so too does evolution and devolution is within *civilizations*.

For example, when one speaks with some tribes located in the western or eastern hemisphere, one understands that in their past they once had grand civilizations. They have legends that they preserve in their memories that are related to their ancestors, very ancient civilizations that have already disappeared. They speak about their ancestors with a lot of respect. Even the primitive-looking cannibals have great traditions: they preserve traditions of immemorial times, of colossal cities, etc. Therefore, they are not "primitive," they are just degenerated people (some very cruel, bloodthirsty and savage tribes are devolutions or descendants of ancient civilizations). Indeed, today it is very difficult to find truly primitive people, because the human race is constantly evolving and devolving.

I repeat: before these five continents existed, Atlantis existed.

In this day and age, we are fascinated with this modern civilization; it is a wonder to us. We wonder about the space machines that travel to the Moon, to Jupiter, or to Venus. Its atomic experiments surprise us, the physiological investigations, the study of living cells, etc. We are so fascinated with these experiments that, indeed, we believe that this is the most powerful civilization that has ever existed in the world. We have fallen into a kind of "geocentric system" similar to those of ancient times. In the Middle Ages, as you very well know, people used to believe that all the stars rotated around the Earth. Once again we have fallen into a kind of "geocentricism" since we believe that the history of the world has orbit around our much boasted civilization.

I think that a "modern geocentricism" is necessary, a new Newton who will be able to demonstrate that our much boasted about civilization is just one of the many civilizations that have existed here on the planet Earth.

The day will come when it will be possible to prove all of this.

There are systems that make it possible to verify the fact that prior to our civilization (which looks so bright) there was a more powerful civilization. Emphatically, I am referring to systems that study the Akashic Archives of nature, the memory of nature.

Experiments with carbon-14, for example, have demonstrated that the Moon is older than the Earth. We can also demonstrate it through systems that make it possible to read the memories of nature. The Akashic Archives are a reality; we do not deny that one day scientists will find them. We, the Gnostics, however, have methods that make possible the study of the Akashic Archives of nature.

Whosoever wants to study those Akashic Archives will have to extraordinarily develop the lotus of a thousand petals (Sahasrara chakra, which is related to the pineal gland) as well as the latent powers that exist within the pituitary gland (the lotus of two petals and ninety-six radiations). These two small glands are extraordinary. Once these glands are developed, they allow access to the *ultra*, to extrasensory perceptions, and also to the Akashic Archives of nature.

When one studies the Akashic Archives of nature, one can study them as if one is watching a "living movie." The whole history of Earth and its root races are recorded in the Akashic Archives.

The sages who have studied the Akashic Archives know that Atlantis was real, that it was an enormous continent that spread from south to north. That gigantic continent was the scenario of the root race that existed before us. I am referring to the great Atlantean root race.

That root race was a race of giants. That is why ancient legends talk about Briareos, the one with one hundred arms, from a race of real cyclops. That root race had a powerful civilization, millions of times more powerful than ours.

Regarding the science of transplants, they were capable of transplanting organs of any type: the liver, the kidneys, the heart. Formidably, they even did brain transplants. In the field of nuclear physics, they attained a massive system of atomic illumination. All their cities used atomic illumination: their farms, their houses, were all illuminated by nuclear energy.

In the field of mechanics, I can state that their cars were not only amphibious, but they could also fly and were propelled by nuclear energy.

They extracted nuclear energy not only from uranium and radium but also from many other metals, and even from vegetables (which was also a cheaper method).

In aerodynamics, they had airplanes more powerful than those of present times: real flying ships propelled by nuclear energy.

They performed better journeys to the Moon, better than those presently performed by this present civilization. They had amazing atomic spaceships in which they traveled to the Moon. Atlantean astronauts descended onto the Moon and, moreover, they also visited the other planets of our solar system. Therefore, we do not even come close to them with our highly boasted about civilization and modern pseudo-wisdom.

In reference to anatomy and biology, they made discoveries that we do not even remotely fathom.

Katebet (Jezebel), "the one of the gloomy destinies," was an Atlantean queen who lived youthfully for thousands of years. Unfortunately, and this is how the decadency of the Atlantean civilization began, she established dismal cannibalism. At that time, girls, young men, etc., were sacrificed to their gods. The young, sacrificed cadavers were transported to the laboratory where the needed glands were removed and used to replace the already wasted glands of the famous Jezebel (the one of "gloomy destinies"). However not only did they remove the physical glands from those cadavers, but also something else.

Today, the famous scientists are so degenerated that they do not know how to manage the principles of life. Yet,

Atlantean sages knew how to use the vital principles of the endocrine glands. They did not ignore that the vibrations of the ether, or in other words the tattvas, enter into the endocrine glands (the small micro laboratories that produce hormones) and never exit.

Atlanteans knew how to use those tattvas or vibrations of the universal ether. Therefore when a gland was transplanted into Jezebel they also used the tattvas. This is why those scientists were absolutely superior to modern endocrinologists, who know nothing about this. Modern endocrinologists ignore the existence of the tattvas because they have never even tried to study Rama Prasad or Dr. Krumm-Heller.

There was a marvelous Atlantean university. I want to emphatically refer to the Akaldan Society, a true university of sages. They studied the law of the eternal Heptaparaparshinokh (the law of seven). They learned how to concentrate the solar rays and make them pass through different chambers. They knew how to transform the seven colors of the solar prism, how to get the "positive," that is to say a "slide" of the solar prism rays.[2]

It is one thing to see the seven colors of the prism and another to transform them into a positive form, to get the positive.

The modern scientists have studied the seven fundamental colors of the solar prism but they have not obtained a slide of those seven colors. The Atlantean sages knew how to get the real positive from the seven colors of the solar prism, and with that positive of the seven colors, they made real prodigies.

I remember the case of two Chinese sages that performed experiments (in the Atlantean style) with the seven colors of the solar prism. They extracted the positive from the seven colors. For example, they put opium in front of a red ray, and then they saw how the opium was transformed into other substances. They took a piece of bamboo and saturated it with a certain substance of a blue color (using the positive part of the solar prism). Afterwards, they saw how that piece of bamboo was intensely dyed with that blue color.

---

2   The author is using the terms for modern slide film to explain this technology.

They played, for example, the sound of the notes do or re or mi in combination with a certain color. Afterwards, they saw how the note had transformed the color; the note transformed it into a completely different color.

So, the law of the Heptaparaparshinokh was profoundly studied in the Atlantean continent. The seven rays were used in their positive form and they made prodigies.

A sage took the milk of a nanny goat and mixed it with the resin of pine on a plate of marble; he later saw that after its decomposition the substance had formed seven different layers. This experiment induced the study of the law of the eternal Heptaparaparshinokh, the law of seven.

Within the field of science the Atlanteans attained extraordinary things. The scientists were also extrasensory clairvoyant savants. They could manufacture a robot that had the intelligent principle of a vegetable or of an animal (the elemental consciousness) and that principle functioned as the consciousness, soul, or Spirit within the robot. So, those robots were real living creatures who obeyed their masters and lords.

The Atlantean root race existed before this present root race. They had titanic cities, but unfortunately they degenerated. They made the atomic bomb and even deadlier weapons. Entire cities were devastated during their wars; many cities were converted into holocausts, atomic holocausts.

Therefore, if we believe that we are the wisest race in the entire universe, then we are absolutely wrong because before us existed a more powerful, a more civilized, a more cultured race. Indeed, in comparison to them we look like uncivilized and uncultured barbarians.

It is unfortunate that Atlantis degenerated. However, we have to understand that all races are born, grow, develop, and die!

Horrible things happened during the Atlantean decadency. Obviously Atlanteans degenerated by means of despicable vices, homosexuality, lesbianism, drugs, etc. They abused everything during the time of their degeneration. Obviously, this is why that race had to be destroyed.

Did that humanity have seven sub-races? Nobody can deny that, but in the end, they degenerated.

The sages of the Akaldan Society performed notable experiments. They knew about the mystery of the Sphinx. They were the first to use it by placing it in front of their university. Much later, when those sages of the Akaldan Society understood that the great catastrophe was approaching, they immigrated to a small continent called Grabonksi. Later, the new lands that emerged from the oceans were joined to Grabonksi, which thus formed the continent of Africa, which is now much bigger.

In the beginning, the members of the Akaldan Society went to the south of Africa. Afterwards, they immigrated to Cairona (today Cairo city). In the land of Nivea (the Nile or Egypt) they established their famous university. They placed the Sphinx in front of it.

The symbol of the Sphinx has the claws of a lion, which represents the element fire, the human head represents water, the hooves of the bull represent the element earth, and the wings represent the air. These are in relation with the four necessary virtues in order to achieve the realization of the Inner Self, the Being. One has to have the courage of the lion, the intelligence of the human being, the wings of the spirit, and the tenacity of the bull. Only in this way is it possible to achieve the realization of the Inner Self, the Being.

The Akaldan Society in Cairona (today Cairo city) established a temple of astrology. The stars were not studied with telescopes as they are today; instead, they were studied with the sixth sense.

When the pyramids are examined (especially the Great Pyramid), it is possible to see some kind of "tubes." These tubes are canals that run from the bottom (from the profound underground sepulcher) all the way to the upper part of the pyramid (to the superior part). Much has been speculated about or stated

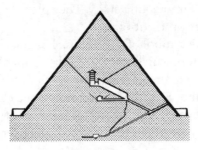

about these "canals." Indeed, these canals
are telescopes. The observatory was not
in the upper part of the pyramid, but
underground in the bottom part of the
sepulcher. Inside, at the bottom of the
pyramid, they would put a receptacle
with water. On a pre-determined day,
when they knew that a star would be visible through that canal, the star was then
reflected upon the water.

The Atlantean adepts of astrology
observed the studied star on the water not
only with their physical faculties but also
with their psychic ones. Instead of observing the sky, they observed the water. Thus,
there in the water they studied the stars
with their sixth sense.

The brothers and sisters of the Akaldan
Society, those great sages, were very good
astrologers. When a child was born, a
horoscope was immediately done. Not the
present modern style of horoscopes, the
present conventional type; these were very
different horoscopes. The wise astrologers observed the stars directly with
procedures that are unknown in this day
and age. This is how they were capable
of reading the horoscopes of the chil-

dren. Obviously, they never failed in their prophecies and calculations.

The newly born children were immediately married. They already knew who each child's spouse would be. We do not want to state that they had to live together from the beginning as husband and wife, because that would be absurd. We want to state that they knew who was going to be the spouse of the newborn child. Thus, at the right moment, that person was informed who was going to be his/her spouse. Once they reached adulthood, they were married.

Thus, with mathematical precision, under the direction of their astrologers, Atlantean citizens were oriented in their vocation or profession. Atlanteans knew what each citizen was born for, what each human being was useful for, because each human being is useful for something. What mattered is to know what a human being is useful for. So, those sages knew the skill of each creature that was born. The sages of the Akaldan Society never failed!

Before the earthquakes and tidal waves shook the continent, they migrated out of Atlantis. They moved in time, because they knew very well that the end was near.

Of course, when the revolution of the Earth's axis came, the poles were transformed into the equator and the equator into the poles. Yet, the Atlantean astrologers had already been warned; they knew that the time was near for that continent to be submerged into the bottom of the dark ocean. It was then that the splendidly dressed crowds entered into their temples. One of those temples was Ra-Mu's temple.

> "In the crowd, people were dressed splendidly, wearing jewels. Their cries and moans filled the air. The people went to find a hiding place in the temples and in the castles. So the wise Mu, great priest of Ra-Mu, stood up and said, 'Didn't I predict all this to come?'
>
> "Men and women, dressed with their precious clothes, covered with precious stones, begged, 'Mu, save us!'
>
> "Mu answered, 'You will die with your slaves and your treasures. From your ashes new people will come. If these people forget that they should not amass material things not only

for their own progress, but also for the generosity towards mankind, the same fate will surprise them.'"

Tradition states that the flames and the fog choked the words of Mu. Within a few months, the country and its inhabitants were smashed and swallowed in the abysms of the ocean.

The words of Ra-Mu were useless, since it is stated that the smoke and the flames suffocated his last words. Atlantis was submerged with all its millions of inhabitants. Today, entire palaces are at the bottom of the ocean and are the habitats of seals and fish. Entire cities are submerged at the bottom of the Atlantic Ocean. That gigantic continent, bigger than all America from Canada to Chile and Argentina, disappeared. It was a huge continent with a powerful civilization!

Therefore, dear reader, our present civilization is not anything special. The present civilization is not the first nor will it be the last. The present civilization is not the highest of all nor will it be the most extraordinary. Indeed, until now it has been the poorest, the most degenerated.

Can we presently conquer space? Are we capable of traveling in atomic spaceships to Mars, Mercury, or Venus? Is there even a project? Yes, there can exist beautiful projects, but are we presently doing anything?

Regarding transplants, is it already possible to transplant brains?

Are we already capable of creating robots with intelligent principles?

There is nothing of that; there is no reason to believe that we are the most powerful civilization. Indeed, this boastful modern civilization with its vulgar generation will also disappear, and from it not even a stone will remain upon another! Our modern civilization, that is to say, "Babylon the great, mother of all fornications and abominations of Earth" will be destroyed very soon.

We feel proud of our supersonic planes and we believe that we are the masters of creation, but soon nothing will remain, absolutely nothing of this perverse civilization of vipers!

So before the existence of this root race that lives on the five continents, the Atlantean root race existed. The descendants of the Atlanteans are the Mayans. The Mayans immigrated to Tibet, Egypt, and Central America. It seems incredible, but even now, the Mayan language is still spoken in Tibet; it is a sacred, ritual language in that country. Let us remember that the Naga and Mayan languages are very similar.

Jesus of Nazareth learned Mayan in Tibet. For example, the sentence pronounced by Jesus, "Eloi, Eloi, Lama, Sabachthani" that some state signifies, "Lord, Lord, how you have glorified me" and others state signifies, "Lord, Lord, why have you forsaken me?" is not, indeed, a Hebrew phrase. That is why when the Jews listened to Christ saying, "Eloi, Eloi, Lama, Sabachthani," they said to themselves, "This man calls for Elias to come to save him..." But any small Indian of Yucatan, Mexico, and Guatemala can translate the sentence, "Eloi, Eloi, Lama Sabachthani," because it is the Mayan language and not Hebrew. That is why the Jews, I repeat, did not understand it. According to the Mayans and their translation this phrase means, "I hide myself in the pre-dawn of thine existence." It is a Mayan ritual sentence.

The Turanians were also survivors of Atlantis; unfortunately, they were devoted to black magic. They also reached Tibet (as other descendants did, such as the selected Aryans), and immigrated in the direction of ancient Persia. The great law finally defeated them and they were all destroyed.

The red-skinned natives of North America are descendants of Atlantis. The ancestors of Mexico, the ancient Nahuatls-Zapotecs, Toltecs, etc., came originally from Atlantis. Almost all the tribes of America descended from Atlantis.

When we advance in these studies we understand that our present root race did not have its origin, as many believe, on the continents on which we live. We comprehend that we come from another root race, the Atlantean root race.

We do not come from apes (orang-utans, monkeys) as Mr. Darwin and his partisans mistakenly believe. We descended, I repeat, from the Atlantean root race. This is clear.

However, the Atlanteans with all their powerful civilizations did not descend from the Atlantean continent. Atlantis, with all its powerful civilizations was magnificent, but the Atlanteans did not descend from Atlantis: they descended from **Lemuria**.

Lemuria was even older than the Atlantean continent. The Lemurians lived on a continent that existed in the Pacific Ocean. It was a continent that existed in that boisterous ocean. It was a colossal continent that covered almost all of the Pacific Ocean. It was bigger than Atlantis, bigger than Europe, bigger than Asia.

Obviously, the Lemurian civilization was powerful. The Lemurians were a root race of giants, of cyclops. The height of the Lemurians was about four to six meters. They were giants; it was a root race of giants.

Lemuria had a powerful civilization, titanic, formidable. In Lemuria, enormous cities were built, cyclopean, surrounded with stone walls and lava of volcanoes. Like now, many people also lived in the country.

In the beginning, in the pre-Lemurian era, we can state that a race of hermaphrodites existed, they were Lemurian hermaphrodites. The division into different sexes happened in the post-Lemurian era. Therefore, we can divide Lemuria in two halves or times: the first part was the existence of hermaphrodites and the second was the division of the race into two sexes.

Let us observe that Lemurian human root race. In the beginning, separate sexes did not exist, the race was hermaphroditic. At that time, each sacred Lemurian had both sexual organs (masculine and feminine) totally developed. Thus, reproduction was performed through the system of **gemmation**. The hermaphrodite detached from its ovaries (through menstruation at a specific time) an ovule or egg perfectly developed. It had the size of a chicken egg with its calcareous shell. When that egg was put in a special environment, it gestated a new creature within its interior. Finally, when that creature was out of the shell, he was then nourished with the breasts of the father-mother. That is how the Lemurians used to reproduce themselves in the beginning. The sexual act did not exist because each individual was complete. Reproduction was through the gemmation system.

However, it so happened that when the post-Lemurian era came, some children were clearly born with one sexual organ more accentuated than the other. Some children were born with the masculine organ more developed than the feminine, or vice versa. That phenonema became more and more common to the point that finally unisexual children, man and woman, were born. Yet, it did not happen from one day to the next, this process of division into different sexes took millions of years. That is why it is written:

> "Eve was taken from one of the ribs of Adam." —Genesis 2:21-22

Adam and Eve are a symbol to represent the division into different sexes.

When the complete division of sexes appeared it was necessary to have sexual cooperation in order to create. Menstruation remained in the feminine sex, but that ovule was unfertilized. It was then necessary to have the cooperation of the masculine sex to fertilize the ovule in order to reproduce the species.

The Creators, the Elohim, the Kumaras, used to congregate the people in order for them to reproduce during pre-determinated times of the year. It was amazing how those races, those tribes, travelled from one point to another in order to go to the temples in which they would reproduce themselves.

The sexual act was never performed outside of the temple. The sexual act as a sexual sacrament was only performed within the temple. Intercourse was a sacrament of the temple. In the paved courtyards of the temple, the couples (man and woman) were sexually connected in order to procreate under the direction of the Kumaras.

The Lemurian humanity enjoyed spiritual faculties. They could perfectly perceive all of the wonders of nature and the cosmos. Their capacity of vision allowed them to see half of a holtapamnas (half of the tonalities of the universal color). We know very well that a holtapamnas has five and a half million tonalities of color.

Their hearing was sensible; they could listen to the symphonies of the universe. Their sense of smell was so sensitive that it was more acute than that of today.

In their alphabet, the Lemurian humanity used 51 vowels and 300 consonants. The power of the word, of the verb, was not degenerated. Their language was the universal language that has power over fire, over air, over the waters, and the earth.

The Lemurian humanity was a superior humanity, millions of times superior to this present humanity. They made powerful civilizations, and also knew how to use the energy of the atom and cosmic rays. They had spaceships in which they traveled throughout the infinite space. These were marvelous spaceships.

Any human being in Lemuria could live for twelve to fifteen centuries. Their lifespan was more than one thousand years. It was a strong, vigorous race. They could easily take a boulder and throw it far away, a boulder that today would require a strong crane to move (perhaps not even a crane would be able).

The Lemurians were a very strong, vigorous root race. However, their origin was not in the Pacific, as it is believed. The ancestors of Lemurians came from the Hyperborean continent, that horseshoe shaped land close around the North and South Poles.

In the Hyperborean continent there existed a powerful androgynous race. They were not hermaphrodites, but androgynous. It was not a root race that was only limited to the terrestrial layer as the Lemurian race was. The Hyperboreans were different: they floated within the atmosphere, within the atmosphere of those days.

However, they created their own civilization. Many have believed that they never knew war, but as a matter of fact there was a race of Hyperboreans that had wars.

The mineral, vegetable, animal, and human kingdoms were very mixed then. There were mineral-vegetables and vegetable-minerals, animal-vegetaloids, and vegetaloid-animals. Human beings were totally androgynous; they could elongate their bodies voluntarily to a great height or reduce themselves to a mathematical point.

The Hyperboreans reproduced themselves as corals do: by **sprouting**. We know very well that there are plants that can reproduce themselves by simple sprouting. A sprout germinates, and this grows and develops. Likewise, from those bodies a sprout could be born that would later detached itself when ready for independent existence, and it would become a creature that was fed by the father-mother.

It was a root race of very combative, tall, and thin beings. They protected themselves with big shields and lances. They used to fight amongst their tribes.

The Hyperboreans lived in a very different era of the world's history. They had their spiritual vision totally developed. Their pineal gland was protuberant; this gland allowed them to see the *ultra* of everything.

If we comprehend that any plant is the physical body of a plant elemental, then we can understand that each plant has a soul [consciousness]. The soul of each plant is its plant elemental. So, the Hyperboreans did not see a forest as we do today, as a group of trees or something like that. The forest to them was endowed with elemental giants with big arms moving from right to left (such as Briareos, the one with one hundred arms). The forest was not silent either, since all around them they could listen to the voices of those colossuses or elemental giants, the voices of the elementals of the gigantic trees.

They had another way of observing things, not as we do now, with our miserable sight that can only perceive physical things. When developed, the sight can allow one to perceive the superior dimensions of nature and the cosmos. It was an omniscient sight. With such sight one could see the Earth as it is, and not as it apparently is, not as we observe it now.

They had a superior wisdom and knowledge, immensely superior to what we presently have. All that we know now is only good in order to organize the intellect a little bit, that is it.

The Hyperboreans were wiser and were governed by the super-humans, by the superhumans of all times and ages.

They had kingdoms and civilizations, but their racial origin did not originate in the Hyperborean continent either. They knew that their ancestors went back further in time. The Hyperborean's ancestors were the protoplasmic humans, the polar humans, the glacial humans of the first root race who lived in the North Pole region.

One cannot avoid laughing at Haeckel's protoplasmic theory and his partisans who believed that life comes from it, which is in accordance with their unbreakable dogma of evolution (this was also accepted by Darwin and his partisans).

The protoplasmic humans were not related to that protoplasm (of some other authors) "that was floating in the ocean." No indeed, they were not related with that protoplasm. It is better if we think about the protoplasmic human and about the protoplasmic root race that existed on the sacred island (the first island, and the last one that will disappear). I am referring to the Nordic land, the Land of Crystal, as our ancestors of Anahuac called it. That island is the ancient Tule, the continent that now is covered by the ice of the North Pole. Yet, in that era that continent was located at the equatorial zone of the Earth, because the Earth's position was different. At that time the current equator was a pole and the poles were the equator.

That island had huge and profound forests. A gigantic polar civilization was created on it. The Earth indeed had a magnificently beautiful blue color. The mountains were as transparent as crystals.

The protoplasmic root race reproduced themselves with the system that we still have in the blood of our organism: the system of **cellular division**. We very well know that this process begins when the germinal cell is divided in two and then four and so on during the nine months of gestation. The germinal cell is divided in two, the two are divided in four, the four into eight, and in that way starts the process of gestation through cellular division. Even now, I repeat, that process still exists in our organism. Why does it exist? It exists because it existed before, because the polar humans reproduced themselves with that process. Thus, at a certain time the father-mother organism was divided into two, as the cell divides itself, thus in that way reproduction was performed. When a creature was born it was celebrated as a great event.

The hierophants had meetings in the temples in order to work with the elements of nature and the esoteric symbols were used in different forms. They were used to indicate to us that life was in the process of materializing into the physical.

The human beings of the polar era could voluntarily stretch or reduce their bodies to a mathematical point. They were androgynous. They were able to express the feminine aspect within themselves, resembling beautiful ladies, as well as being able to submerge within themselves and express their masculine aspect.

They were authentic, divine, androgynous beings. Their imagination reflected the starry firmament; they spoke the word of gold that flows like a river through a thick, sunny jungle.

Then, Uriel, great master of Venus, taught the arts and sciences. Uriel left a book written with Runic characters, a book that was studied by the human beings of the polar root race, the human beings of the first era.

The whole of this data is written in the Akashic Archives of nature. If you develop the epiphysis and the hypophysis (those two small glands) you will be able to study (with proper concentration) for yourselves what I am stating here.

What is the origin of the polar root race? The human beings of the first root race knew quite well that they had evolved in a previous era, or that they had lived within a superior dimension, in the fourth dimension; they knew that it was there that they had lived and known the mysteries of the universe.

The human beings of the fourth dimension did not ignore that they had come from the fifth dimension. The human beings of the fifth dimension did not ignore that they had come from the sixth dimension, and also the human beings of the sixth dimension never doubted that they had unfolded from the primitive original seed.

Therefore, the elemental seed, atomic, primeval of the human race, existed before the existence of these universes; it came from within the Chaos.

The seeds of the human race, of the plant elements and of the animal species were within the Chaos. Those seeds were asleep within the Chaos before the universe existed. When the universe was shaken with the Word, when the Creative Word of the first moment put in movement all of the atoms, those seeds surged forth from within the Chaos. They had their first manifestation in the seventh dimension. Afterwards they crystallized and developed a bit more in the sixth, then in the fifth, later in the fourth, and the day came when those seeds appeared already with some development on our planet Earth, on the protoplasmic Earth, as simple living protoplasm.

The human race came from the Chaos, evolved in the Chaos, and was developed within the Chaos. One day, the human organisms will return to their seminal primeval state, thus they will come back into the Chaos. They came from the Chaos and to the Chaos they will return.

Once our planet Earth was a protoplasm, yet in a future time, after the seventh root race, it will be a cadaver, a new moon. Then life will pass, it will exist within the superior spheres and it will eventually return into the Chaos, because from the Chaos it came and to the Chaos it will have to return.

Does the Chaos that I have mentioned have any relation with the theories of Darwin, Haeckel, Huxley, and their partisans? No, it does not have any relation with those nonsensical theories. The Chaos is the Chaos and the reason for its existence is the Chaos itself. The Chaos is sacred. The latent seeds of life exist within the Chaos, there they developed and from there they evolve and descend, from dimension to dimension, to finally appear here in a concrete form.

I wrote this book in order to talk in detail about this. However, there are other authors that have explained many aspects of anthropogenesis. I can especially recommend the second volume of *The Secret*

Doctrine, entitled *Anthropogenesis,* whose author is the Master H. P. Blavatsky. Also Rudolf Steiner, for example, in his *Treatise of Occult Science,* clarified many aspects about it.

From the Chaos emerged the cosmos. Undoubtedly, through the law of three, through the Holy Triamazikamno, it is possible for the creation of new unities. When the three forces (positive, negative, and neutral) coincide at a common point, a new creation appears. The creation of any cosmic unity would not be possible without the conjunction of these three forces that in themselves are the Holy Triamazikamno. These three forces are the Holy Affirmation, the Holy Denial, and the Holy Conciliation.

However, to create is one thing and to organize is another. It is possible to create, but if there is no organization, what could the creation itself be worth? For a cosmos (an order of worlds) to emerge, to crystallize, it is necessary to have another law. I want to emphatically refer to the law of the eternal Heptaparaparshinokh, the law of seven. Through the law of the Holy Triamazikamno creation is done, but through the law of seven the organization of what has been created (in the form of a cosmos) is established.

Therefore, our solar system exists because of two laws: first the Holy Triamazikamno; second the eternal Heptaparaparshinokh. Thanks to these two laws our solar system and planet Earth can exist. From the Chaos emerged a cosmos, and from the Chaos emerged all the cosmoses. So, from the darkness emanates the light...

Chapter One

The materialistic anthropologists of this decadent and tenebrous age have been investigating the origin of the human being, yet indeed all that they have elaborated upon are **hypotheses**.

If we ask the anthropologists of conventional anthropology exactly when and how the first human being appeared, they would indeed not know how to give us an exact answer.

Since the times of Darwin and Haeckel and up to this present day and age, innumerable theories about the origin of the human being have appeared. However, we must emphatically clarify that none of these much boasted about theories can be demonstrated by themselves. Ernest Haeckel himself emphatically asseverated that within the field of conventional science neither geology nor that science named phylogeny will ever be exact. Therefore, if Haeckel himself made this type of assertion, what more can be added to this subject-matter?

Indeed, this matter about the origin of life and the human being certainly cannot be known as long as humanity does not study Gnostic anthropology in depth.

What do the materialists that study protists say? What do they so arrogantly affirm? What do they suppose about the origin of life and of the human psyche? Let us remember with complete, absolute clarity Haeckel's famous atomic Monera that was within an aqueous abyss, a complex atom that could not, in any way, emerge by chance, as this good gentleman supposed. Haeckel (an ignorant in depth) was worshipped by the British people. He induced great damage onto the world with his famous theories. As a parody to Job, all we can say about him is the following: "May the memory of him perish from the earth, and may he have no name in the streets." [Job 18:17]

Do you believe perhaps that such an atom from that aqueous abyss, the atomic Monera, could emerge by chance? If the intelligence of scientists is necessary in order to construct an atomic bomb, then how much more talent would be required in order to construct an atom?

Therefore, if we deny the intelligent principles of Nature, then it's mechanism would cease to exist, since the existence of

Nature's mechanism is not possible without machinists. If someone considers it possible for a machine to exist without its inventor, I would like him to demonstrate it. I would like him to place the chemical elements upon the table of the laboratory in order for a radio or an automobile or simply an organic cell to emerge by chance.

At this time, we know that Don Alfonso Herrera (the author of Plasmogeny) managed to build an artificial cell, yet this cell was always a dead cell; it never had life.

What else do the materialists that study protists [microorganisms] say? They say that the consciousness, the Being, the Soul, the Spirit, or simply the psychic principles, are nothing more than the molecular evolution of the protoplasm throughout the centuries. Obviously, the molecular souls of these fanatic materialists that study protists will never endure a deep analysis.

Therefore, the soul-cell, Haeckel's famous gelatinous Bathybius from which all organic species emerged, is indeed just a good subject-matter for Molière and his caricatures.

What lies at the bottom of this whole subject-matter and what is behind all of these mechanist's and evolutionist's theories is the impulse to combat the clergy. They are looking for a system that satisfies the mind and the heart in order to demolish the Hebraic *Genesis*. It is precisely a reaction against a misunderstanding of the biblical Adam and his famous Eve (who was made out of one of his ribs).

Therefore, this *reaction* against a *misinterpretation* of the biblical Adam and Eve is the source for the ignorant theories of Darwin, Haeckel and their other accomplices. So, it is not right to originate so many hypotheses (that in themselves are deprived of any serious foundation) because of mechanical reactions against misapprehensions.

What does Darwin state about the matter of the catarrhine monkey? That the human being possibly came from it? Nonetheless, he does **not** emphatically asseverate it, as the German and British materialists supposed. Indeed, Charles Robert Darwin placed within his system a certain basis that disagreed with and even absolutely annihilated the supposed human emergence from the monkey, whether it is the catarrhine or platyrrhine.

First of all, as Thomas Henry Huxley already demonstrated, the human skeleton is completely distinct in its structure from the skeleton of the monkey. We do not doubt that there is a certain similitude between the anthropoid and the wretched intellectual animal mistakenly called "human being," however this resemblance is not definitive or defining on this matter. The anthropoid has a climbing skeleton. It is made for scaling. This is what the elasticity and construction of its skeletal system indicates. On the other hand, the human skeleton is in itself made for walking. Definitely, these are two totally different skeletal constructions.

Chimpanzee and human skeletons compared

Moreover, the flexibility of the bones of the cranial axle of the anthropoid and of the human being is completely different; this invites us to seriously ponder. On the other hand the following has been stated by the materialistic anthropologists with complete, absolute clarity: an organized being cannot in any way be the outcome of another being who marches in the opposite way, who is antithetically ordered.

We must give a certain example in reference to this matter: Let us observe the human being and the anthropoid. Even though the human being in this day and age is certainly degenerated, he still is an organized being. Now let us study the life and behavior of the anthropoid; we can observe that it is organized in a different way, contrary, antithetical. Therefore, an organized being cannot be the outcome of another who is organized in the opposite manner. This former assertion is always severely uttered by the materialistic schools.

Which age could be associated with the anthropoid? In which epoch did the first simians appear upon the face of the Earth? Unquestionably, who can deny that it was during the Miocene Epoch? Obviously, it had to have appeared during the late Miocene Epoch, 15 to 25 million years ago.

Why did the anthropoids have to appear upon the face of the Earth? Can the people associated with materialistic anthropology, those brilliant modern scientists who boast about being so wise, give us an exact answer...? They obviously cannot.

Moreover, the Miocene Epoch was not in any way located upon the famous Pangaea which is touted so much by the materialistic geologists. It is obvious that the Miocene Epoch had its proper scenario on the ancient Lemurian land, the continent that was formerly located in the Pacific Ocean. Remnants of Lemuria

Easter Island

are still located in Oceania, in the great Australia, and on Easter Island (where some carved monoliths were found), etc. So, if the materialistic doctrine cannot accept this due to the fact of their narrow-mindedness that is bottled up within the idea of Pangaea, what does it matter to people, or to science, or to us? Indeed, they are not going to detect Lemuria with carbon-14 tests or with potassium argon or with pollen. All of these test systems of a materialistic type are just good materials for Molière and his comedies.

In this day and age, after the infinite suppositions made by Haeckel, Darwin, Huxley and all of their secularists, they continue enthroning the theory of natural selection (of the species), granting it nothing less than the power to create new species.

In the name of the truth, we must state that natural selection as a creative power is simply a rhetorical game for ignorant people, something that has no basis.

The assertion that states that new species are being created through natural selection, that the human being had emerged through natural selection, is in the depth frightfully ludicrous and shows ignorance taken to extremes.

We do not deny natural selection; it is obvious that it exists, yet it does not have the power of creating new species. The truth is that physiological selection, selection of structures, and the segregation of the most apt does exist, that is all.

To take natural selection up to the degree of converting it into a universal creative power is the breaking point of absurdities. A true sage would not have so stubbornly conceived of such a notion.

Never have we observed or witnessed a new species emerge through natural selection; if so, when? In which epoch?

Structures are selected, yes, we do not deny it. The strongest ones triumph in the struggle for daily bread, in the incessant battle of every moment, when one fights in order to eat and not be eaten. Obviously the strongest one triumphs, and he transmits onto his descendants his characteristics, his physiological particularities, his structural particularities. Thus the selected ones, the most capable, are segregated and transmit their aptitudes on to their descendants. This is how the law of natural selection must be understood; this is how it must be comprehended.

Any given species within the profound jungles of nature has to fight in order to devour and not be devoured. Logically, such a struggle is frightful. As an outcome, as is proper and natural, the most powerful ones triumph. There are marvelous structures within the strongest, and their important characteristics are transmitted onto their descendants. Yet, this does not signify a change of figure; this does not signify the birth of a new species.

So, never has a materialistic scientist observed one species emerging from another through the law of natural selection. This has not been proven, this has never been palpated in any way. Then what do these materialistic scientists base themselves upon? It is easy to throw a hypothesis out there and then emphatically asseverate that it is the truth and nothing but the truth. Nevertheless, aren't the scientists from materialistic anthropology the ones who state that they believe in only what they can see? The ones who do not accept anything that they have not examined? So, they contradict themselves horribly, because they believe in their assumptions that which they have never seen or touched.

They affirm that the human being comes from the mouse, yet this is not proven; they have never perceived this directly. They also emphasize the idea that the human being comes from the baboon or the mandrill (*quadrumana cynocephalus* from western Africa). The sophisms of these foolish scientists are innumerable, absurd assertions of facts that they have never seen.

We, the Gnostics, do not accept their superstitions because their absurd assertions are fetishism. We, the Gnostics, are mathematicians when investigating and precise in our expressions. We do not like such fantasies; we want facts, concrete and definite facts.

Thus, when investigating this theme related with our possible ancestors, we can clearly verify the chaotic state in which the materialistic doctrine is found. The disorder of these scientists' degenerated minds and their lack of capacity for investigating is evident. This is the crude reality of the facts.

The subject-matter that states that certain hominoid forms emerge from other ones, just like that, based only upon ridiculous tests (such as carbon-14, potassium or pollen) palpably constitutes the shame of this century.

We, the Gnostic anthropologists, have different systems for investigation; we possess special disciplines which allow us to put into activity certain latent faculties of the human brain, certain senses of perception completely unknown to materialistic anthropology.

It is logical that nature has a memory; one day this will be demonstrated. Scientific research has already started; soon the sounding waves will be rearranged into images which will be perceptible on certain screens. Certain technical experiments about this matter already exist, then the tele-viewers of the whole world will see the origin of the human being, the history of the Earth and of its root races. When that day arrives (a day that is not so distant), the Antichrist of false science will be naked before the solemn verdict of public consciousness.

Indeed, the problems of natural selection, the climate, the environment, etc., fascinate many people, and this is why they forget the original source from which each species emerged.

The stubborn scientists believe that natural selection can be processed in an absolutely mechanical way without intelligent directrix principles. This is as absurd as to think that any machine in the world can be processed without an intelligent principle, without an architectural mind or without an engineer to give form to it.

Undoubtedly, these intelligent principles from nature can only be rejected by the stubborn ones, for those who pretend that it is possible for any organic machine to emerge by chance. These principles could never be rejected by those truly wise humans in the most complete sense of the word.

As time passes by and as we delve into all of this, we can see and further find all of the mistakes in materialistic anthropology. Therefore, it is necessary to profoundly reflect on all of these things.

If instead of assuming a position of attack against the clergy, those materialistic anthropologists would first of all reflectively analyze in depth, then they would not dare to disseminate their anti-scientific hypotheses.

We know that Adam and Eve — who bother the scientists from materialistic anthropology very much— are nothing more than a symbol. So, the profane scientists from materialistic anthropology want to refute the Biblical *Genesis*. It is good for them as well as for all of us to understand that the book of *Genesis* is just a treatise of alchemy for alchemists. *Genesis* must be studied as an alchemical book, and not in a simple, literal way.

So, the scientists from materialistic anthropology exert themselves in order to refute something that they not even remotely know. This is why I honestly dare to state that their hypotheses do not have solid foundations.

Darwin himself never thought to go so far with his doctrine. Let us remember that he himself speaks about the characterizations: Unquestionably, after some organic species have passed through a selective process of physiological structures, these species characterize themselves in a constant and definitive way. Then, we see that the famous anthropoid had to pass through selective processes and subsequently assumed all of its present characteristics; but, it did not pass through any change again, this is obvious.

That subject-matter about Pithecus-Noah with his three famous sons — namely: the cyanocephalus with a tail, the monkey without a tail, and the paleolithical arboreal man — indeed never had exact corroboration; these are just theories without any basis, and indeed they are frightfully ludicrous.

Those who, when inquiring about the origin of the human being incline themselves towards the prosimian mammals (for example, the famous lemur), show that they do not even remotely suspect what the human being himself is, let alone his origin.

In this day and age, the celebrated lemur is considered by some scientists as one of our conspicuous ancestors because of his assumed discoid placenta.[1] This has nothing to do with human genesis. This, in depth, is nothing but fantasies that are deprived of any reality.

Many renowned materialistic scientists are active in order to study the mechanical evolution of the human species or of any other species. They ignore that the original microorganisms from this great nature, human beings or beasts, always develop themselves within psychological space and the superior dimensions before crystallizing themselves in a physical

1   Because the shape of the placenta in both humans and Lemurs is disc-like, scientists proposed that the two species were related.

form. Halfway through their development these microorganisms crystallize into a sensible form; they had previously passed through tremendous evolving and devolving processes within psychological space, within the hypersensible, within the superior dimensions of nature and the cosmos.

Of course, when we speak like this, conventional anthropologists feel very nervous and upset. They feel like the native Chinese people do when listening to an occidental musical concerto. Perhaps they laugh, ignoring that, "The one who laughs at what he does not know is an ignoramus who walks on the path of idiocy." [Goethe]

Materialistic scientists search for similarities. Yes, they believe that the shape of the shark's head and the mouth gives origin to other mammals, among them our "brother" the mouse that for the materialistic scientists in this day and age has passed into the category of "great lord." Supposedly, the mouse is nothing more than the ancestor of those such as Haeckel, Darwin and possibly of Huxley and Einstein, even of those famous pharaohs of ancient Egypt, and who knows what else.

So, in this day and age, the mouse is considered a prosimian mammal. The mouse has become a prominent subject-matter in the conference halls. Alas, behold how far the ignorance of the human being has gone.

We do not deny that the mouse existed in Atlantis and that indeed it was the size of a pig. Don Mario Roso de Luna, the eminent Spanish writer, clearly talks about this. The small Larousse Dictionary illustrates it, and states that in ancient times the mouse was denominated with the word "alto" (tall).

Yes, the mouse existed in Atlantis and we cannot deny its presence in Lemuria either. Yet, to assert that the mouse is one of the most important ancestors of the human being is totally false. Indeed, when one does not know Gnostic anthropology, one then falls into the most frightful absurdities.

Now, in this present Space Age, the accomplices of the Antichrist, materialistic science, bow before the mouse and the shark (which is also considered an ancient ancestor) and also before the lemur, which indeed is a very fascinating animal.

So, when Gnostic anthropology is known in depth, one cannot fall into ludicrous statements; this is logical. When the principles of materialistic anthropologists are carefully analyzed, we discover that their fantasies are due to their absolute ignorance of universal Gnosticism.

It is very empirical to establish a basis for a possible descendant of the human being based upon the fact that the feature of one face is similar to that of another. This is too shallow; so too are those who suppose that the human being was made out of clay; they are not aware that this is merely a symbol.

So, as we have already stated, before the original microorganisms of this great Nature crystallized into a physical form such as into human beings or beasts, they always first developed themselves within psychological space and the superior dimensions.

There is no doubt that these microorganisms are similar in their construction, therefore they could not serve as basis, as a foundation in order to establish a theory or simply in order to elaborate a basic concept. These microorganisms are differentiated from one other by the pace at which they crystallize. This is normal indeed.

So, the origin of the human being is something more profound. He developed himself within the Chaos, in the superior dimensions of Nature, until in ancient times he was crystallized into a tangible form.

Unquestionably, in future chapters we will advance more and more throughout all of this exegesis.

I want to state to you that the origin of humanity will be uncovered in these lectures, namely the primary and secondary causes that originated the human species, and other themes of transcendental repercussion.

Do the materialistic anthropologists know perhaps the answer to the former questions? The scientists themselves, who are followers of Haeckel, know very well that the whole geological past and materialistic phylogeny will never be exact sciences. This is what they have affirmed themselves; this is how they have stated it. If this is what they say, then what?

We are in moments of great inquietudes. The mystery of the origin of the human being must be clarified. The field of conjectures is detestable; it is like a wall without a foundation: all you need to do is to push it a little bit in order for it to become a pile of rubble.

The most critical aspect of materialistic anthropology is the denial of the intelligent principles of universal machinery. Obviously, such an attitude leaves the machinery without a foundation. It is not possible for the machinery to work or to be built by chance. The intelligent principles of Nature are active in the whole selective process and they manifest themselves wisely.

Likewise, it is an absurdity to bottle up ourselves within the dogma of mechanical evolution. If the *constructive* principles exist in Nature, unquestionably, so too do the *destructive* principles.

If there is evolution in every living species, likewise so too is devolution.

For instance, there is evolution in the seed that dies in order for the stalk to sprout. There is evolution in the plant that grows, that gives flowers, leaves and that finally gives fruits. Yet, there is devo-

lution in the plant that withers, that ago-
nizes, and that finally becomes a bunch of
wood.

There is evolution in the creature that
is gestated in the maternal womb, in the
child who plays, and in the teenager. Yet,
there is devolution in the elder who dwin-
dles away and finally dies.

When the worlds emerge from the
Chaos of life they start to evolve to a certain point; afterwards, they devolve and
finally they become new moons.

If when we study anthropology we give
exclusivity to mechanical evolution, then
we are studying it in a partial way and
thus fall into error. However, if we study
anthropology in the light of the law of
devolution, then we advance equilibrated,
because evolution and devolution constitute the two laws of the mechanical axis
of the whole of Nature.

It is a total absurdity to say that evolution is the sole foundation for this entire,
great natural mechanism. We must consider life and death, the times of development and the times of decrement. Only
in this way will we correctly march within
the integral structure of the Gnostic dialectic.

We are not willing, in any way, to remain bottled up within the materialistic dogma of evolution. We must by necessity study the devolving processes of anthropology, otherwise we will walk on the path of error.

What are the original prototypes of this human race? Who knows them? By means of scientific methods we can see, hear, and touch the original prototypes of this human race. We know very well that human beings existed before the intellectual animal appeared in the Atlantis of Plato. Atlantis is not a simple fantasy, as the ignorant fanatics of the famous materialistic Pangaea suppose.

The human being existed in Lemuria, as well as in the Hyperborean and Polar epochs. However, these are themes that we will develop in the following chapters in order for us to clarify these anthropological themes better.

Atlantis really existed. The archipelago of Antilles, the Canary Islands and even Spain itself are remnants of Atlantis. Spain is a piece of ancient Atlantis.

This is unknown to those who are passionately fond of materialistic anthropology and also to the geologists who are so deeply behind the times, who are incapable of projecting themselves through time. How can they know something about the events that occurred many millions of years ago in the Miocene Epoch? What do they know about it? Have they seen it? Have they touched it?

If we Gnostics talk about the Miocene Epoch, it is because we can see it. The Miocene Epoch is accessible to the one who is capable of developing the faculties that are latent in the human brain. Nonetheless, the materialist attitude of denial is incongruent. They say that they only believe in what they see, nothing else, yet they believe in all of their absurd fantasies. No one has seen and no one has proven their suppositions.

We can state with certainty that not even one scientist has ever witnessed how the first human being came forth. Yet, they talk with so much self-sufficiency it is as if they were present in the Miocene Epoch, as if they had seen the anthropoids appearing there in ancient Lemuria.

The materialistic anthropologists enthrone their marvelous gods, namely the lemurs and mandrills (baboons); they place them as sublime prosimians from which they assert we descended. Has this been confirmed by them? Have they, at some time, seen it? Never! So, where is their foundation? Is their foundation resting upon fantasies that they have never seen? Are they not the same people who state they only believe in what they see? Then, why do they believe in what they have never observed? Is it not perhaps a contradiction? Are they not perhaps contradictory within their depth?

Chapter Two

The theme of the origin of the human being is indeed very debatable and very thorny.

Charles Robert Darwin left certain principles in his work that must be remembered by the materialistic anthropologists. Darwin states that a species that evolves positively cannot in any way descend from another that evolves negatively. Darwin also affirms that two similar species, even though different, can refer to a common ancestor, but one can never come from the other.

Thus, as we advance through these disquisitions of scientific anthropology, we obviously find certain contradictions within materialism. How is it possible that Darwin's former principles are ignored? How is it possible in this day and age that there are still people who think that the human being comes from the ape?

Unquestionably, the facts speak for themselves, and until now the famous "missing link" has not been found. Where is it?

Much has been stated against the existence of the father Manu,[2] the "Dhyan Chohan," yet indeed there are millions of people in the eastern and western world who accept it. Moreover, such a belief is more logical than Haeckel's monkey-man, who indeed is nothing more than one of his fantasies.

Time is passing by and the famous simian-man has not yet been discovered anywhere on Earth. Where in the world is the ape that is capable of reasoning, thinking, and that has a language? Where is it? Hence, in depth these types of materialistic bookish fantasies are absolutely good for nothing.

Observe for instance the size of the brains: the cephalic mass of a gorilla is not even a third of the brain of any savage person from our terrestrial globe. A link that would connect the most evolved gorilla with the most uncivilized savage person from Australia is missing. Where is that link, what happened with it? Does it exist?

Undoubtedly, the first simians appeared on the Lemurian continent

---

2   From Hindu mythology, the progenitor and lawgiver of the human race. See glossary entry.

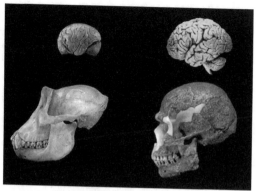

Gorilla and human skulls compared

during the Mesozoic Epoch, but where was their origin? Gnostic anthropology emphatically affirms that certain degenerated groups of humans in Lemuria sexually mixed themselves with beasts in order to originate the simian species.

Haeckel never in any way was opposed to the concept that apes had their birth in Australia, in Lemuria. He always accepted the reality of the Lemurian continent. Yet, let us reflect a little. Where was Lemuria located? In the Pacific Ocean, this is obvious. It covered an extensive zone of that sea. Through ten thousand years of earthquakes, Lemuria was submerged little by little in the boisterous waves of that

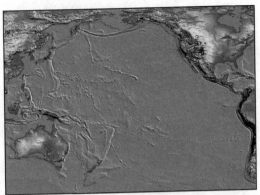

The Pacific Ocean

ocean. Yet, remnants of Lemuria were left, namely Oceania, Australia, Easter Island, etc.

Lemuria was real; it had its place in a very ancient time. This assertion can upset the materialistic anthropologist partisans of Pangaea. These ladies and gentlemen have their minds bottled up within that dogma related to Pangaea; they do not even remotely accept the possibility of Lemuria.

There is absolutely nothing strange about the fact that the simians were born during the Mesozoic Era and evolved to the Miocene Epoch, yet others from the Eocene Epoch, during the Tertiary Period.

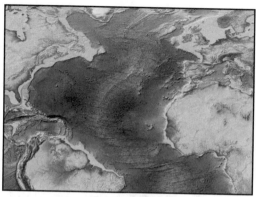

The Atlantic Ocean

However, our assertions do not end here. Other species of apes had their origin in the Atlantis of Plato, a continent that is nothing more than a myth to the materialistic fanatic partisans of Pangaea. Nonetheless, even if they deny it, Atlantis existed. Even if they are opposed to it, Atlantis has already been discovered. Anyone who has studied the marine floor knows very well that between America and Europe there is a great marine platform.

It was precisely a few years ago when several scientists — those who had discovered Atlantis — proposed to explore the remains of Atlantis from the coasts of Spain. However, Spain was ruled by Franco's regimen, therefore they were

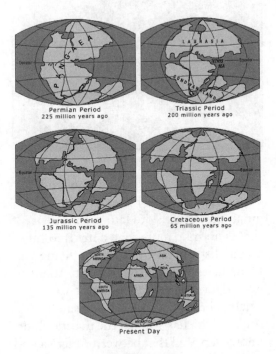

The Pangaean Theory

According to the materialistic "Theory of Continental Drift," the supercontinent "Pangaea" began to break up about 225-200 million years ago. The theory holds that initially two massive continents were formed, eventually fragmenting further to form today's continents. This theory is commonly called "The Theory of Plate Tectonics."

not allowed to perform their investigations. Therefore, Atlantis is not what it is believed to be, that is to say, it is not a fantastic legend, but rather a tremendous reality.

In other times, the map of the world was completely different.

Everything is changing; even the supporters of Alfred Wegener's Pangaea have suffered great changes. We know very well that the floating continents experience some displacements; Mario Roso de Luna had already clearly explained this and therefore it should not surprise anybody. The materialistic fanatics of Pangaea agree with this, they do not deny it. However, they lack a lot of necessary information in order to know the causes of the displacements of these continental flotations. I think that if they would study Mario Roso de Luna, they would better complete their information.

If we compare our Earth to a cracked-open egg, then the yolk would be the continents floating over the egg white. There may be many substances, liquids, and elements between the continent-like yolk and the egg white, yet in this day and age materialistic science does not know anything about them.

There are some who believe that certain types of superior apes come from Lemuria, for example:

· The gorilla, a genus of anthropomorphic ape that comes from Equatorial Africa, and has a stature of about two meters and a maximum weight of 250 kg.

· The orangutan that comes from the Malayan language and means, "a man from the trees," is also known as the great anthropomorphic ape from Sumatra and Borneo; it measures about 1.20 to 1.50 meters in stature; they are arboreal and easy to domesticate.

· The chimpanzee is an anthropomorphic ape from Africa.

There are others who clearly affirm that the inferior types of apes, like the catarrhine or platyrrhine, etc., properly came from Atlantis. We cannot make objections about this, but indeed we must profoundly reflect upon it.

In this day and age materialistic science is making certain very silly statements;

everyday they invent new hypotheses.
So, a chain has been established that is
very curious and ludicrous in content,
a chain that is related with our possible
ancestors. The shark appears to be the
king of that chain, from which, according to those anthropologists, descended
the alligators. Such a theory is ludicrous
in itself; indeed it is only conceivable by
alligator-like minds. They continue stating that the famous opossum comes after
the alligator; they state that the opossum
is a creature similar to a crocodile but
a little more evolved (this is what they
emphasize). And on they go, following
the course of that great chain of marvels
to a certain little animal. In this day and
age they have given a lot of importance to
that little animal; I am emphatically referring to the lemur. They grant unto the
lemur a discoid placenta, a matter that is
rejected by zoologists.

We find gigantic contradictions within
these gorges of false science. They proceed
by stating that from the lemur (which
could have existed some 150 million years
ago) descended the monkey, and finally
the gorilla appears at the end of this
chain. In this fantastic chain, the gorilla is

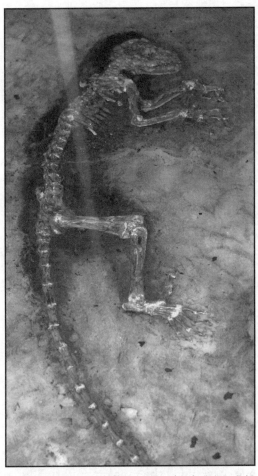

Darwinius masillae, a type of lemur proposed as the "missing link" between prosimian and simian ("anthropoid") primate lineages. This is the only fossil ever found of this animal, even though it is proposed as the origin of the entire human race.

our immediate ancestor, the forefather of the human being.

As I have already stated in the first chapter, some anthropologists do not set aside the wretched mouse while talking about these themes; they even want to include it in their fancy chain. In which link of this ludicrous chain do they place the mouse? Between which animals should "Mr. Mouse" be? Who knows? Let them suffer with their theories. Based on the fact that the mouse is small, they affirm with a tone of extraordinary sapience that the human being was also so minute, so microscopic, that is to say, so tiny, that in this day and age we would be overwhelmed if we were to see such a human being. On what do they base their theories in order to asseverate such nonsense?

So, should we accept that according to them human beings are children of the mouse? I do not know in which link of that chain they include the mouse, perhaps before or after the lemur.

They sustain that we grew until we reached the present human height of this great civilization, and until we reached the "extraordinary perfection" that this civilization possesses now.

So in this day and age, the mouse is the priority in public anthropological conferences of this great civilization. Thus, if things continue in this way, then within a short time the governments should prohibit the killing of mice, since according to the materialistic anthropologists the mice are our forefathers.

Where are all the links of that chain? How is it possible that a shark turns into an alligator, from one night to the next, or throughout many centuries, just like that? Many millions of years have passed and the sharks continue to be sharks. Never has it been seen that from one species of sharks, whether from the Atlantic or Pacific Oceans, new alligators were born. To begin with, the crocodiles, alligators, or caimans — at least the ones we know — are not found in the seas, but rather in rivers and lakes, except the crocodile-like anthropologists who are very civilized and who are going around the world inventing silly theories.

Do you know of a species of alligators that has emerged from within the boisterous waves of the ocean? The world knows

very well that the habitat for all types of
alligators is the sweet waters.

We have seen alligators in great rivers
and this is easily proven. We have visited
the oceans and we have never seen or
heard of any fisherman that has caught
an alligator within his net in the middle
of the ocean. They might catch sharks,
but alligators? When?

We are asserting things based on clear,
concrete, and definitive facts.

Where might the links be that bond the
alligator to the opossum? Where might
the links be that bond the opossum to
that small animal the lemur, an animal which is deprived of a placenta and
because of this Haeckel labeled it a creature with a discoid placenta? Continuing,
where are the links that bond the lemur
with the monkey? Where are the links
that bond the monkey with the gorilla?
And where are the links of the gorilla to
the human being? Where are they? We
have seen precise elements and we observe
that the points of connection are missing.

To state things just like that is abundantly absurd. Much has been said about
the Monera, an atom of the aqueous
abyss, the first drop of salt within a siluri-

an ocean which was filled with mud at its bottom and where the first bed of rocks was not yet deposited. However, what is the origin of the Monera? Could we perhaps conceive that something so extraordinary as the first atomic point of protoplasm which is so properly organized and with such a complex construction can be the outcome of *fortuity*, the outcome of *chance*?

We understand that when we deny the intelligent principles of nature, then the whole organization of protoplasm doesn't make any sense.

Time is passing by and materialistic anthropology is going to be destroyed little by little. The materialistic anthropologists still cannot say when and how the first human being appeared. They state nothing more than ludicrous hypotheses, conjectures that lack serious foundations.

Materialistic anthropology points to Australia a lot. The position of conventional anthropology is highly supported when they state that the tribes that live in Australia descended from apes. Scientifically, the former statement falls to the floor when we measure the size of the brains and confront the facts. We

see that the size of the brain of a highly
developed gorilla does not even reach half
of the mass of the brain of an uncivilized
native of Australia. Therefore, a point of
union between the two is missing.

So, where is that link? Let them show
it to us, for we are all here to see it. In my
first lecture (the first chapter) I said that
the ladies and gentlemen of materialistic
anthropology affirm in an extremely gran-
diloquent way that they only believe in
what they can see, yet the facts are showing us their falsity. They firmly believe in
absurd hypotheses which they have never
witnessed. The fact of attributing or stating that human beings come from sharks,
to establish a whimsical chain just based
on morphologic resemblance, shows us
in the depth a superficiality taken to
extremes. If what they write about is nonsense, they are indeed abusing the readers'
intelligence a lot. If what they talk and
teach about is nonsense, then they are
indeed terribly comic and even ludicrous.

Human beings were sexually mixed
with beasts in Lemuria and this is something we have no doubt about at all.
Apes were not the only outcome of this
mixture, so were various monstrous

forms that even in this day and age have documentation in the Eastern part of the world as well as in the Western part.

We will cite as an example certain estranged Lemurian simians that could serve as a mockery for the superficial materialists of this epoch. Yet, one has to affirm the truth with valor. These Lemurian simians were a certain species that existed; they were soon walking using their hands and feet as any simian would; before long they were standing on their two feet. They had blue and also red faces. These simians were the outcome of certain human beings who crossed themselves with subhuman beasts of the Miocene Epoch, especially from the Mesozoic Era. We find everything that we formerly stated within reference materials such as on papyrus, codices, bricks, ancient monuments, and archaic manuscripts. Thus, there were multiple simian forms that emerged on the ancient continent of Mu.

Nevertheless, how did the human being first appear? In which way did it occur? Until now, all of these interrogatives have been a puzzle, a true enigma for the materialistic partisans of Darwin, Haeckel, and even for modern anthropologists.

Where can we find the origin of the human being? Unquestionably, we can find it *within the human being himself.* Where else could it be?

Let us now focus our attention on Australia. What do materialistic anthropologists asseverate? They affirm that the Australian Aboriginal tribes have the anthropoids as their ascendants. Of course, they cannot prove this, yet they affirm it, they believe in this. See for yourselves how paradoxical these ladies and gentlemen truly are.

The Australian clans are the most primitive that presently exist in the world. What could the origin of such families be? We must first begin with the origin of Australia. Australia is a fragment of Lemuria, which was located in the Pacific Ocean. It is an old land.

Where are the ancestors of those tribes? Let us talk about their physical bodies. We will obviously find such osseous remnants at the very bottom of the Pacific Ocean. There we can find skeletons of beasts, since the Australian clans that presently exist are a mixture of certain animal humanoids that passed through many transformations. Such families should be observed, thus we can understand that they are the outcome of the sexual crossing of the inhabitants of ancient Lemuria with the beasts of nature.

The bodies of some of the native people from many parts of Australia are so abundantly hairy that it seems that they have the skin of beasts. The former statement is the apparent foundation for those materialistic ladies and gentlemen to asseverate: "Lo and behold...these are children of anthropoids... our theory is proven!" Those materialistic anthropologists are extremely superficial; they do not have any maturity in their understanding. Regrettably, their minds are degenerated, in a state of decrepitude.

If we want to search for the origin of the human being we must know ontogeny in depth. It is not possible to know phylogeny if we exclude ontogeny.

Observe the recapitulative processes of the human being inside the maternal womb.

Nature always makes recapitulations. Observe a seed, the germ of a tree: inside it is a potential tree. All it needs is to be developed. So, in order for it to develop, it needs soil, water, air, and sunlight. All of the processes passed on by the tree that served as a parent are recapitulated by nature within the seed that must be developed. Using other terms, we will state that all the processes through which the whole family of that tree endured over time are recapitulated by nature within the seed that must be developed. The whole species of that tree developed slowly and grew at the same pace of the other trees; likewise, so did the tree from which that seed was detached. In itself that seed repeats that process of recapitulation, leaf by leaf, until the tree finally bears its fruits and seeds so that the other trees that will eventually be born from them will make the same recapitulations.

Let us observe how nature recapitulates all of her marvels within the cosmos. Each year spring, summer, fall, and winter return; these are perfect recapitulations.

Thus, within the maternal womb is a correct recapitulation of the whole human species. All the steps the human being passed through since its most remote origins are repeated inside the maternal womb.

First of all, no one can deny that inside the maternal womb the fetus passes through the four kingdoms of nature. First it is a stone, second a plant, third an animal, and fourth a human being.

As a human embryo or corpuscle it is inorganic; it is the ovum that detaches itself from the ovary and goes to join the organic matter. Thus, circulation carries the ovum to the very bottom of the matter for its own development.

In its second aspect we see how the human embryo assumes a vegetal state. The human fetus looks like a carrot, rounded at its base and pointed in its superior part. When this is clinically studied, it seems much more like an onion with distinct layers; between the layers there is

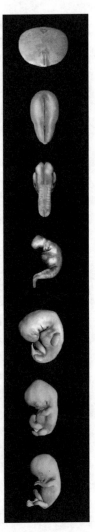

a marvelous liquid. The possibility of the fetus depends upon the umbilical cord of this apparent onion (just like the fruit of any plant). Behold here the vegetal state.

Later on the animal form appears, then the fetus takes the form of a tadpole, and this is already completely proven.

Finally, it will assume the human form.

So, the four phases — one mineral, one vegetal, one animal, and the last one human — get their recapitulation.

After knowing and analyzing all of the former phases, we conclude by stating the following: no physician has ever seen the form of the anthropoid in any of the four phases of the fetus. During the process of fetal recapitulation, which scientist has ever observed the fetus taking the form of a catarrhine or platyrrhine monkey or the shape of an orangutan or a gorilla? Therefore, what materialistic science affirms is absurd.

Therefore, the origin of the human being has to be found within the very womb of the woman. The origin of the human being is hidden within the processes of recapitulation; these are the stages though which human beings have passed.

So within a human mother's womb, in which process of recapitulation can we see the appearance of the shape of a shark or that of a lemur? Where does all that Haeckel mentioned so passionately appear? At which stage of the pregnancy does it appear? Why do these materialistic ladies and gentlemen want to stray from what is correct? Why don't they search for the origin of the human being within the human being itself? Why are they searching for the human being's origin outside of it? All the laws of nature are inside of us, and if we do not find them within, we will never find them without.

We have then arrived to a very delicate and extremely difficult point. To state that we were a stone, a plant, an animal, and finally a human has all been very well settled. But when and how did it happen? Which primary or secondary causes are governed by all of those processes? These are enigmas. If the ladies and gentlemen of materialistic anthropology were not so mesmerized by the dogma of Euclid's three-dimensional geometry, then these processes could be easily clarified and everything would be different. Unfortunately, on their own whim they want all of us to accept what they state as a dogma of faith. They want to have all of us bottled up within that foundation. This is absurd, as absurd as to want to bottle universal life or as absurd as to want to pour an ocean into a glass.

The fourth coordinate, the fourth vertical, exists, this is undeniable, yet this statement upsets the materialists. Nonetheless, Einstein, who cooperated with the manufacturing of the atomic bomb, accepted the **fourth dimension**. Nobody can deny the fourth dimension in mathematics. Yet, the materialistic people of this day and age cannot accept the fourth dimension, not even when it has been mathematically proven that other superior dimensions exist within nature. They want to be forcefully enclosed within Euclid's tridimensional world. This is why physics is totally stagnant in its advancement due to their absurd disposition. This is the time in which there should be cosmic ships capable of traveling throughout the infinite, yet that longing is not possible while physics continues to be bottled up in Euclid's three dimensional dogma.

These ladies and gentlemen have not been capable of answering the question of how and when — on which date — the origin of the human being occurred. So, if they would accept the possibility of the fourth, fifth, sixth and seventh dimensions, then everything would be different. Yet, we are sure that they will never accept this. Why? It is because their minds are in the process of complete degeneration, due to abuse of sex. Therefore, under these conditions it is not possible for them to comprehend the thesis that we are stating here. If they want to understand it, they need to begin to regenerate their brain; thus, this will be the way for them to accept our Gnostic postulations.

Stone, plant, animal, and human being: behold here the foundation for serious anthropology. Now, let us think about the former state of the human, about our legitimate ancestors. Unquestionably, we will find them in animal life in nature but located in the fourth dimension. This statement is distressing for the materialists. Nonetheless, they are the same materialists who were laughing at Louis Pasteur and his theories, who mocked him when he was disinfecting surgical utensils. They never believed in microorganisms because they were not able to see them, yet now they accept them.

Can there be animal life within the fourth coordinate? Yes, of course there is. Could there be a system of corroboration? Obviously yes, yet the methods are very different from those of false science, which is found in a retarded state. Who has those processes and systems? We have them and will we gladly teach them to everyone who truly wants to investigate the field of pure science.

Did animal life exist within the fourth coordinate? Yes, it logically did. Did plant life exist in the fifth coordinate? Yes, naturally it did. Did mineral life exist in the sixth coordinate? Yes, it did. Yet, I clarify that the mineral life of the sixth coordinate, the plant life of the fifth, and the animal life of the fourth were not in any way similar to the animal, plant, and mineral life of this merely physical world. The animal, plant, and mineral life of the fourth, fifth, and sixth coordinates condensed here in this tridimensional terrestrial globe, we do not deny it, yet this happened over the course of millions of years.

How can we in any way define the evolving processes if we eliminate from nature that subject-matter stated by Gottfried Wilhelm Leibniz? We are referring to the **Monads** or **Jivas**, intelligent principles of nature. Certainly, a whole abyss interposes between the atomic Monera of Haeckel and the Zaristripa of Manu, the Jiva of the Hindustani or the Monad of Leibniz. This is because the atomic Monera of Haeckel is very far from the true Monad or principle of life.

It is true, no lie, for certain, that the virginal sparks or, simply put, Leibniz's Monads, evolved within the mineral kingdom during the epoch of great activity of the sixth dimension. These Monads evolved likewise within the vegetal kingdom in the fifth dimension, and finally they advanced to the animal state in the fourth dimension. This is unquestionable.

These dimensions from nature will be seen in the future through the use of some devices of high optical pressure. Yet, until that day arrives, we can be sure that we, the Gnostic anthropologists, must tolerate the same mockery Pasteur endured when he was talking about microbes. However, the moment will arrive in which those dimensions will be perceptible by means of televisions; then the satires will end.

Meanwhile, as we speak experiments are being performed in order to try to transform sound waves into images. When this occurs, all of the evolving and devolving processes of nature will be seen. Then the Antichrist of false science will be exposed to the solemn verdict of public consciousness.

In regards to the human organism, we see that in the beginning it is invisible; that is, when the process of gestation begins, when the primeval cell begins to germinate, the ovum and sperm cannot be seen with a simple glance. Who would suppose that a creature could emerge from one spermatozoon and one fertilized cell? Can this be seen by a simple glance? Thanks to the microscope we know that it is so; this is obvious.

Thus, by establishing facts we can state that the Monads who passed through the mineral kingdom in the sixth dimension are the same ones who passed through the plant kingdom in the fifth dimension, and through the animal kingdom in the fourth dimension. It was precisely at the end of the fourth dimension when a certain creature similar to an anthropoid (from the Greek *anthropoeides:* "with a human shape") appeared. Nonetheless, this creature was not a gorilla, a chimpanzee, or anything of the sort.

When the epoch of activity for this three dimensional world started, that creature — as well as the planet Earth — suffered some changes and various metamorphoses. Finally, the human figure crystallized.

We have to take into account that with the passing of the centuries the morphology of all human creatures (as well as of nature) is changing. Unquestionably, human morphology emerged in accordance with the Protoplasmic Age of our Earth in order to really come into existence. Then, afterwards it passed through the Hyperborean, Lemurian, and Atlantean periods, alternating itself a little until reaching our present day.

The creatures, the human beings, the ancient human races that preceded us, were giants who slowly over time lessened in stature until becoming what we presently are; this is all stated by traditions of ancient Mexico and by other countries on Earth.

We will continue by explaining about the four steps — mineral, plant, animal, and human — exclusively within the tridimensional zone of Euclid on this physical planet Earth.

So, materialistic anthropology failed to give us the how, when, and why the first human being appeared. After having delved very deeply into these matters, I know for sure, I am absolutely positive that for the materialistic anthropologists all of these statements will become new enigmas without any solution.

Therefore, at the present time, the scientists have no other choice but to accept the crude reality of the superior dimensions of nature and the cosmos. They may deny them, since they have the right to deny them. They might laugh, however we have already stated many times, "The one who laughs at what he does not know is an ignoramus who walks on the path of idiocy."

So, as time passes, materialistic science will become exposed by new discoveries, and each day it will sink more and more in the pit of its own ignorance.

Indeed, the theory of Pithecus-Noah with his three bastard sons, namely the cyanocephalus with a tail, the monkey without a tail, and the arboreal man, are very good material for Molière and his caricatures. Indeed, in our blood we do not have anything of pithecoids. For us, the facts speak for themselves.

Chapter Three

The hour has arrived for us to make a certain analysis in relation to the origin of the human being. But, indeed, in the name of the truth, we have to state that the merely materialistic anthropology knows nothing about the origin of the human being.

Nevertheless, we Gnostics know the origin of the human being. In our former chapters, based on our Gnostic anthropology, we taught some basic knowledge related with its origin. Yet now we are going to delve a little more into this enigma.

Let us think for a while about the Mesozoic Era of our world, on the era of the reptiles [66 to 252 million years ago]. The human being indeed existed at that time; of course, this is denied by materialistic anthropology. Indeed, the merely profane anthropology ignores the real origin of the human being.

Materialistic anthropology wants us to believe that the human being did not exist before the Quaternary Period [the most recent 2.6 million years].[3] The possibility that the human being existed during the Cenozoic Era [from today back 65 million years] is denied to him. In depth, this is manifestly absurd.

Nonetheless, some facts make us think: we ask ourselves why certain species such as the plesiosaur and the pterodactyl survived for such a long period of time and then finally became extinct; now, in this day and age all we find from these species are fossil remains located in certain museums.

Meanwhile, in spite of the disappearance of these species, in spite of the fact that they were extinguished from the face of the earth, the human being still continues to exist. Why did almost all of the species from the Mesozoic Epoch become extinct? How is it that human beings were not extinct?

So many species have disappeared, yet the human being continues to be alive. How can this be? What explanation could materialistic science give for this phenomenon? Logically, materialistic science cannot give us a single explanation.

---

3   A recent fossil discovery has extended this date to 2.8 million years ago.

Obviously, if many contemporaneous species of the human being from the Quaternary and Tertiary Periods disappeared, then, human beings should also have eclipsed from the face of the Earth; the human species should have also disappeared. Yet, they continue to exist.

Therefore, the fact that the human being did not disappear in the Quaternary and Tertiary Periods allows us to infer that the human being existed long before the Quaternary Period and even beyond the epoch of the reptiles [Mesozoic], or of the Carboniferous Period [299 to 359.2 million years ago].

Moreover, we have the right to discuss the existence of human beings before the Mesozoic Era [66 to 252 million years ago]. that right is granted onto us precisely due to the concrete fact that all of the species from the Tertiary and Quaternary Periods have disappeared, and nonetheless their contemporaries, the human beings, still continue to exist.

Gnostic anthropology, through superior systems, knows that the other species from the Tertiary and Quaternary Periods vanished and that the intellectual animal mistakenly called "human being" did not disappear because the human being existed before the Mesozoic Era and before the Carboniferous Period. These facts are clear to us; facts are facts, and before the facts we have to surrender.

For example, there is a statement about flying serpents; this is terrifically true indeed, and the Bible corroborates it. The Bible cites the Leviathan in the book of Job. The Zohar emphatically affirms that the tempting serpent of Eden was a flying camel.

It is not irrelevant to remember that in Germany[4] a species of flying camel was found; I refer to the fossil remnants that were perfectly organized by anthropologists. It is 78 feet long, a giant; its long neck is similar to that of present day camels and it was provided with membranous wings. When the fossil remnants of this body of these are observed, then it is evident that indeed it is related to a flying serpent (similar to a camel because of its long neck).

These fossil remnants might perhaps belong to the Leviathan. What could the

4   There is also the case of when the first pterosaur fossils were discovered in Germany in 1784, scientists thought they were aquatic animals. It took them 100 years to finally recognize they were flying reptiles!

anthropologists state in relation to this matter? Obviously, this saurus, or its better if we say, pterosaur, is in depth only the remnants of what flying serpents were in the archaic times of our Earth. We will conclude by stating that the pterosaur has more in common with the ophidioid than with the lacertid.

Delving further within all of this, we find many other pieces within the field of anthropology. In the Imperial Library of Peking there were paintings upon which some plesiosaurs and pterodactyls were shown. We ask ourselves: how is it possible that ancient people who did not know anything about paleontology or paleontography knew about extinct species from the epoch of the reptiles?

If we did not know about the possibility of developing certain capacities and faculties of a transcendental type in the human brain, the former paragraphs would be inexplicable. Such faculties allow us to study the history of nature and of the human being, a history that lies hidden within the very depth of the memories of everything that is, has been, and shall be.

Dear readers, for certain we must know that the present human species is not in

any way made up of real human beings —
it is true, no lie.

Gnostic anthropologists are in agree-
ment with profane anthropologists only
on the issue of the origin of the intellectual animal. The intellectual animal comes
from the Quaternary Period or at the end
of the Tertiary Period. This is something
that we cannot deny in any way.

First of all, it is convenient to make
a complete differentiation between a
human being[5] and an **intellectual animal**.[6] The true human being existed
before the Carboniferous Period and
before Mesozoic times.

This true human being lived during the
epoch of the reptiles. Unfortunately, some
authentic human beings degenerated terribly at the end of the Tertiary Period,
during the Miocene. They absurdly mixed
themselves with some beasts of nature, as
we have already stated in previous chapters. Several gigantic simians were the

---

5   A real human being has discarded the animal mind,
and created the soul. The real human being is not controlled by nature, whether internal or external. See glossary for more information.

6   An intellectual animal has a human shape but psychologically is just an animal with an intellect. That is, it
is ruled by instinct, passion, desire, etc. See glossary for
more information.

outcome of that mixture; that species in turn mixed themselves with other subhuman beasts. Thus, the outcomes of all of this were the simians that we know, and the evolution of certain humanoids.

During the Quaternary Period these humanoids continued reproducing themselves incessantly. Subsequently, they continued reproducing themselves incessantly, and they are the humanoids of this epoch in which we live. This present humanity is a mixture of authentic human beings with animals of nature.

Now, we see the difference between the real human beings of the first, second, and third root races and the intellectual animals of the fourth and fifth root race in which we are presently. However, we must not be disappointed. The seed for the human being is still in our sexual glands. There is not a single person who does not carry these human sexual elements, since this is the outcome of the crossing of human beings with animals.

Therefore, the possibility exists for present day people to elevate themselves to a true human state since they carry within their sexual glands those human seeds. Of course, we have to develop those human seeds. Indeed, in order to create the authentic human being within ourselves, we need to know the mysteries of sex.

Unfortunately, materialistic anthropologists believe that they are human beings. They do not know anything about the mysteries of sex. They invent multiple theories about the origin of the human species that are in no way useful. I think that all of the speculations stated by the materialistic anthropologists are causing great psychological damage to people. It is lamentable that materialistic anthropology is corrupting this human race. It is already enormously degenerated, and with such fantasies this humanity continues to degenerate awfully every day.

We Gnostic anthropologists have to severely judge the materialistic anthropologists who state that they only believe in what they see. Nonetheless, they believe in what they never have seen. They believe in absurd utopias, such as the one which states that we are the children of mice, or that the mandrill (the baboon) is the ancestor of those elegant ladies and gentlemen.

We have to search for the origin of this fifth human race, which is this present root race to which we belong. We must find its cradle around Kashmir on the central plateau of Tibet, around the Euxinus, etc.

I do not mean to say that the entire cradle of this present root race had its origin only in those cited regions. Yet, in the name of the truth I have to state that those places of the Earth constitute a very important origin for this human species. I am clearly referring to the people of this fifth root race.

There have been five root races in the world. They correspond to five different epochs. We have already stated that the Protoplasmic [Polar] root race came first, afterwards the Hyperborean root

1: Polar Race

↓

2: Hyperborean Race

↓

3: Lemurian Race

↓

4: Atlantean Race

↓

5: Aryan (present) Race

↓

6: [future race]

↓

7: [future race]

race, thereafter the Lemurian root race, more recently, the Atlantean root race, and finally our Aryan root race.

Throughout these chapters we will summarily develop the history of each root race, yet we will include a complete description of the scenarios upon which each root race was developed.

Now I want to state that the human beings of the first root race existed at the northern polar cap on the Sacred Island. At that time, the northern polar cap, the two poles, were located on the equatorial zone.

Unquestionably, the form of life of that root race was very different from that of our present one. Materialistic anthropology does not know anything about this whole matter. Moreover, our assertions will not in any way agree with the famous Pangaea or great primeval continent. Therefore, when we state such declarations, we expose ourselves to only one thing: the mockery of profane anthropologists.

Indeed, profane anthropologists do not totally know about celestial mechanics. They do not know there is a process of the revolutions of the axis of the Earth.

They think that the Earth has always had the same position in relation to the sun. Thus for this reason they have invented their Pangaea, since the theory of Pangaea is more comfortable for them than the study of astronomy.

The Hyperboreans had their scenario on those horseshoe-shaped lands that go around the North Pole. Obviously, England itself and even Ireland were lands that belonged to the Hyperboreans. Alaska also belonged to that land, since all of those regions form a horseshoe around the northern polar cap, although some even of these parts have been under the waves and have come up again since then.

Later on, Lemuria was in the area that now occupies the Pacific Ocean. It was an enormous continent that covered the whole area of the Pacific Ocean.

Subsequently, Atlantis was in the ocean that bears its name.

Therefore, the physiognomy of our terrestrial globe has changed many times. The world has had five aspects or scenarios upon which its five root races developed.

However, we do not expect that the ladies and gentlemen of materialistic anthropology will accept all of this. Unquestionably, this will be something impossible since they believe that they know everything. Nevertheless, not only do they ignore these things, but worst of all they ignore that they ignore.

Materialistic anthropologists who are partisans of absurdity propose to attack the Biblical *Genesis*. Thus, with their anti-clerical efforts they have invented those speculations that are abundant here, there, and everywhere.

They do not even want to understand what the word **Eden** signifies. *Ed-en* signifies "voluptuousness" ["giving pleasure to the senses"]. Its etymology is explained from a Greek basis. Therefore, Eden signifies "voluptuousness."

Eden is sex itself. The whole Biblical *Genesis* is a work of alchemy that literally has nothing to do with history.

That Eden that in former times was situated in Mesopotamia between the Tigris and Euphrates Rivers much later became the school for Chaldeans and Magis, the Alehim. That Eden seems to be related to the famous Adi-Varsha of ancient

Lemurians and even to the Garden of Hesperides of the Atlantean continent.

Eden is sex itself, yet this will never be accepted by the materialistic anthropologists. Moreover, they will never accept the great sexual mysteries of Chaldea, India, Babylon, Mexico, and Egypt.

In Lemuria, the system of reproduction was by **Kriyashakti**. This was during the Mesozoic Era. The Kriyashakti system existed long before the human race ever fell into animal generation (orgasm). We know very well that the Lemurian root race fell into degeneration in the third part of the Eocene, that is, during the Miocene Epoch.

Indeed, the true human beings of the Mesozoic Epoch reproduced themselves by means of Kriyashakti; this word signifies *willpower* and *intelligence*. Lemurians were true human beings, and their system of reproduction would not be accepted in this day and age by the intellectual animals. This is because the system of reproduction of the true human beings, the Kriyashakti, is a sacred system that would cause laughter and rejection by the materialistic anthropologists. They would feel offended with such a system.

In Lemurian times, sex was considered sacred; the sacred sperm was never ejaculated.[7] The sperm was considered a venerable matter. A spermatozoon escaped during intercourse without orgasm in order to fecundate the womb. Thus, the human race enjoyed wonderful powers, extrasensory faculties, that allowed them to know all of the marvels of the universe and the cosmos. This is why it is stated that they lived in a paradisaical state.

Nevertheless, when the human beings fell into animal generation (orgasm), that is to say, when they started to ejaculate the entity of semen, they then precipitated themselves into devolution. So, Lemurians sexually degenerated themselves in the third part of the Eocene Epoch. These sexually degenerated Lemurians were fallen humans who

---

7   Moses wrote: "A man from whom there is a discharge of semen, shall immerse all his flesh in water, and he shall remain unclean until evening. And any garment or any leather [object] which has semen on it, shall be immersed in water, and shall remain unclean until evening. A woman with whom a man cohabits, whereby there was [a discharge of] semen, they shall immerse in water, and they shall remain unclean until evening." - Leviticus 15:16-18. Compare with: "Whosoever is born of God doth not commit sin; for his seed [semen] remaineth in him: and he cannot sin, because he is born of God." - 1 John 3:9

sexually mixed themselves with beasts of nature; thus, intellectual animals were born from this mixture.

Definitively, the intellectual animal will never accept this system of reproduction by Kriyashakti precisely because of his animal condition. The system of Kriyashakti is not for intellectual animals, it is a system for human beings, since animals and humans are two different kingdoms. Indeed, this is why we would not be surprised if the intellectual animals from materialistic anthropology reject the system of reproduction by Kriyashakti.

Nevertheless, in spite of everything, we have to state that the seeds of the human being are still within the endocrine glands of the intellectual animal. Therefore, it is obvious that if we work with the system of Kriyashakti, we can indeed regenerate the brain and develop within our physiologic, biologic, and psychosomatic nature the true human being, the authentic human. Yet, I restate, the intellectual animal does not like the Kriyashakti system.

Gnosis has diffused the mysteries of sex in many places. It is true that universal Gnosticism has accepted the system of reproduction by Kriyashakti, however it is also true that millions of intellectual animals have rejected it. We cannot blame them since they are indeed intellectual animals who are the outcome of sexual intercourse of various Lemurian humans (who sexually degenerated themselves in the Tertiary Period) who mixed themselves with beasts of nature. How could the outcome of humans and beasts accept a sexual system that does not belong to them? It is impossible! Thus, it is good that we reflect upon this.

Now let us consider another more important theme in order for us to reflect upon it in more depth. Let us see: from where did all the living species emerge? From where did this nature emerge? Why do we have to accept all of the materialistic utopias? Why do we have to accept the dogma of evolution? Why should we live within the world of hypotheses?

The moment has arrived to delve a little further into this subject-matter. In my second chapter I stated that the human species had developed in other dimensions. I also asserted that the materialistic ladies and gentlemen do not accept these superior dimensions. They want, by their own whim, to place us within Euclid's tridimensional dogma. They are like the pigs that always want to remain in the pigsty and do not want to see anything that does not resemble it. Yet, we do not accept dogmas.

To begin, they have no proof regarding all of their beliefs. What they have affirmed, for instance, is that they believe that the human being comes from the ape.

Darwin never, ever stated that the human being came from the ape. Indeed, Darwin only affirmed that the human being and the ape have a common ancestor. Therefore, Darwin only opened a door, that is all.

Charles Robert Darwin always felt nonconformity with the Biblical narrative of creation because in the book of *Genesis* God appears as an executioner for the nonbelievers. Nonetheless, Darwin rejected Karl Marx's offer. Karl Marx offered Darwin the dedication of his book, *Das Kapital,* first English edition. Darwin rejected Karl Marx's offer because he did not want to have his name associated with attacks against religion. Even while in states of extreme doubt, Darwin remained with a speck of faith in divinity. In the complete sense of the word, Darwin was never an atheist because he never denied the existence of a God. He stated in his autobiography that he always felt horror to all of that which can cause suffering. He never agreed with the suffering inflicted upon a slave by his ruler or the suffering inflicted upon an animal by any man; for him such horror was one of the reasons to abandon religion.

Darwin was not a materialist, he was an investigator. He opened a door, that is all. We have to take advantage of that door and delve into the mystery, since until now everything has been hypotheses, as when Haeckel stated — and who indeed emphatically asseverated — that neither geology nor phylogeny would ever be exact sciences.

If theories that are invented one day disappear the next day, if those ladies and gentlemen are affirming what they have never seen, if they are lying is such a way, then we must not and cannot give them any credit. Therefore, we must appeal to what Gnosticism teaches, which is to appeal to the wisdom of the ancient sages.

The fact that the human race developed within other dimensions is something impossible for materialistic science to accept, yet for the Gnostics it is a reality.

The reason ancient sages could tell us about plesiosaurs and reveal to us the different beasts that existed during the epoch of primitive reptiles and periods even beyond that, such as the Carboniferous Period, without ever having studied paleontology and paleontography, without having such jargon within their heads, is because they possessed extraordinary faculties that can be developed; these are dormant within the human brain.

Can materialistic anthropologists affirm that they totally know the human brain already? Obviously, they cannot. Moreover, I affirm that medical science still does not know the human body. They might believe that they know it, but they do not.

Conclusion, what is the origin of this humanity on planet Earth? What is the origin of all of the races of nature, of all which has been, is and shall be? This is the theme upon which we need to reflect in this chapter.

What do the ancient Nahuas tell us about the **Omeyocan**? What is the

Omeyocan? It is stated that within the Omeyocan "there is only wind and darkness"; this is how the Nahuas affirm it. The Omeyocan is also named "Yoalli Ehecatl" because of the wind and darkness. This invites us to reflect.

What would the erudites from Mexico and those in the Eastern world, in Asia, tell us about the Omeyocan?

I once stated what "matter" is, in itself. I said that any form of matter can be destroyed, yet the substance (or seed) of the matter continues in other dimensions. Thus, the Earth-substance or the Earth-seed is deposited in the profound space of the universe, within the unknowable zero dimension. I also affirmed that such an Earth-substance was the **Iliaster**. Eventually the Earth would remain deposited as a seed within the profundity of the infinite space, awaiting a new cosmic manifestation.

Let us remember that when a tree dies, its seeds remain. Within each seed there is the possibility of the development of a new tree. Likewise, when a planet dies, then its seed or its homogenous, insipid, insubstantial, colorless, odorless matter remains deposited within the bosom of the eternal Mother Space.

Such a seed in relation with the one is two. We must not forget that in order to be one, there is the necessity of being two. The one feels itself to be two. Thus, the primeval chaotic Earth, that seed-world deposited in Mother Space, is the Omeyocan. It is a true paradise that during the time of inactivity vibrates with happiness.

The Omeyocan is also named Yoalli Ehecatl due to the wind and darkness and because Ehecatl is the god of cosmic movement, the god of wind. Yoalli Ehecatl, the great cosmic movement of the Omeyocan, is the place where the authentic happiness of the world reigns, profound inexhaustible joy.

There are cosmic days and cosmic nights. When the Earth is in a germinal state, when any given planet is found deposited in a germinal state within the bosom of profound space, it slumbers; yet while being two, they are found to be one. After a certain period of activity, the electrical impulse, the electrical hurricane, makes all of the positive and negative aspects enter into activity. This is why it is stated that within the Omeyocan there is wind and darkness. We do not want to state darkness in the complete sense of the word; this is only an allegorical way of speaking.

Let us remember in the Egyptian mysteries when the priests would approach the neophytes. They would whisper in their ears, "Osiris is a dark God." Yet, Osiris was not dark. The fact of the matter is that the light of the pure Spirit, the light of the great Reality, is darkness to the intellect. This is why it is stated that within the Omeyocan there is only darkness and wind — that is to say, cosmic movement. This is where the uncreated light emerges from and where the universal movement represented by Ehecatl is developed.

Before the manifestation of the Solar Logos, that is, the Multiple Perfect Unity, the infinite quietude whirls about within the Omeyocan. In the sacred land of Anahuac,[8] the Solar Logos was always named Quetzalcoatl. Indubitably, Quetzalcoatl as Logos exists long before any cosmic manifestation.

---

8   The central plateau of Mexico.

The Omeyocan is the navel of the universe that intensely vibrates and palpitates within reciprocal whirls, where the infinitely large bursts into the infinitely small, where the large encounters the small, where the macrocosm encounters the microcosm.

With the dawn of any universe, the electrical hurricane makes all of the atoms palpitate in the form of whirls within the Omeyocan, within the navel of the universe, within the cosmic womb, while being two.

In the Omeyocan, the Tloque Nahuaque is the nocturnal tempest of all possibilities. Because when the electric movement, the electric hurricane, the electric whirl, makes all of those atoms spin within the chaotic matter, all the possibilities of universal life persist. This is how the best authors of cosmogenesis have described it.

The Omeyocan is the lord of the night, the black Tezcatlipoca who, when denying himself, bursts into light and the universe is born that fecundates, that drives Quetzalcoatl, the Solar Logos.

Tezcatlipoca in its feminine aspect represents the moon and God Mother

who is precisely the womb of the world.
This is why it is stated that Tezcatlipoca
bursts into light; the Mother, which as
a matter of fact, was fecundated by the
Logos, swells like a lotus flower and thus
finally this universe is born. It is stated in
Nahuatl that Quetzalcoatl then directs
and drives the universe that emerged into
existence.

The Logos, the Multiple Perfect Unity,
is radical. Yet, he turns himself asunder
into the forty-nine fires in order to work
with this rising universe. Unquestionably,
the one who directs the universe is precisely the Logos, Quetzalcoatl. It is the
cosmic consciousness governing, directing
that which is, has been, and shall be...

I am perfectly sure that conventional
anthropology would not accept this conception of Quetzalcoatl. We do not ignore
that materialistic anthropology rejects
the Logos and that it is against the sacred
Anahuac tradition, that it does not want
anything to do with the wisdom of the
sacred land of Anahuac.

When rejecting Quetzalcoatl as the
true monarch of the universe, materialistic anthropology actually declares itself
against the wisdom of the sacred land

of Anahuac. Thus, my dear friends, it is
worthwhile that we reflect upon this.

Nevertheless, it is not convenient that
we Gnostics have an anthropomorphic
conception of our Lord Quetzalcoatl
either, no! Therefore, I repeat,
Quetzalcoatl is a Perfect Multiple Unity;
it is the Greek Demiurge, the Platonic
Logos. Quetzalcoatl is the prodigious
principle of nature that makes every atom
vibrate, that makes every sun shake; it is
the creative fire of the first instant.

Never could the partisans of materialistic anthropology asseverate that they
know the fire; they do not know it, much
less the electricity. We are interested in
the fire of the fire and in the profound
knowledge of electricity.

They pass judgment on the fire and
they consider that it is an element of combustion, yet they are mistaken. Indeed,
they might state, "If we strike a match,
we will see that fire emerges; so this is the
outcome of combustion." Yet, it is not
so. Rather, combustion is the outcome
of fire, because the hand that strikes the
match has its own fire (energy) in order
to move, and within the match itself there
is latent fire. By means of striking the

matchbox, we remove the particles within which the chemical matter of that match is trapped, so that the fire can emerge. *The fire exists before striking the match;* this is unknown in chemistry. The fire in itself is the Logos, the fundamental intelligent principle of nature.

Let it be clear that we are not defending that anthropomorphic god who upsets the materialists so very much, no! We are only making a great emphasis here in order to state that nature has intelligent principles and that the sum of all of them is Quetzalcoatl, the Greek Demiurge, the Platonic Logos, the Multiple Perfect Unity that is latent in every atom, within every corpuscle that comes into life and within every creature that exists under the sun.

There is no doubt, my dear friends, that monotheism caused a great deal of harm to humanity, because materialism, atheism, was the outcome of it, the consequence of it.

In addition, when polytheism was taken to its extremes, it caused a great deal of harm because from it emerged monotheism, and likewise from monotheism emerged materialistic atheism.

Thus, with polytheism degenerated, it gave origin to anthropomorphic monotheism, and due to the abuses carried out by the diverse religious clergies this as a result brought materialism.

If we accept the intelligent principles within the depths of nature and within the cosmos as the foundation of the whole machinery of relativity, then we will not ignore that "variety is unity."

I understand that in the not so distant future, humanity must return to polytheism, yet in a transcendental monistic way. This must be equilibrated from the spiritual point of view, meaning from the spiritual monotheistic and polytheistic points of view. Only in this manner, indeed, can a renovation of principles and a complete revolution of the consciousness be initiated.

Eon	Era	Period		Epoch
Phanerozoic	Cenozoic	Quaternary		Recent / Holocene
				Pleistocene
		Tertiary	Neo-gene	Pliocene
				Miocene
			Paleogene	Oligocene
				Eocene
				Paleocene
	Mesozoic	Cretaceous		
		Jurassic		
		Triassic		
	Paleozoic	Permian		
		Carboniferous	Pennsylvanian	
			Mississippian	
		Devonian		
		Silurian		
		Ordovician		
		Cambrian		
Precambrian		Protezoic Eon		
		Archean Eon		

Geologic time scale according to modern science.

Chapter Four

First of all, it is good to know that anthropologists speak to us about three very important epochs: first, the Paleozoic Era, second, the Mesozoic Era, and third, the Cenozoic Era.

They emphatically affirm that during the Paleozoic Era the first unicellular beings, the microorganisms, the molluscoids, the mollusks, the fish, and the first reptiles were in the waters of life. This is affirmed with incredible conviction by the materialistic anthropologists as if they were present in those archaic epochs, as if they saw, smelled, touched and even heard all of that which occurred in those ages.

Nonetheless, as I have already stated in previous chapters, the anthropologists from materialism always asseverate that they only believe in what they can see. They state that they will never accept anything that they have not seen with their eyes or touched with their hands. Although I have reiterated this statement many times, I have to repeat that such an affirmation is absurd, completely false, since they believe in what they have never seen. They affirm in an absolutist way their false suppositions.

When did they see the Primary Epoch? When were they present during the Paleozoic Era? Did they live during the Mesozoic Era? Did they per chance exist during the Cenozoic Era? They affirm mere hypotheses that they have never witnessed. They affirm what they have never seen. Nonetheless, they believe themselves to be eminently practical.

They have never seen the Paleozoic Era. What do they know about it? What do they know of the life forms that existed in that first age, and about the events of that archaic epoch of our world?

They also talk about the Mesozoic Era, the epoch of the great antediluvian reptiles. We do not deny that the reptiles existed on the Earth. It is clear that the epoch of reptiles existed, that is a fact. Our Earth was populated by enormous reptiles, this is undeniable. Let us remember the brontosaurus or bronsaurus, the plesiosaurus, the pterodactyl, etc.

All of those were indeed gigantic, enormous monsters that were one or even two street blocks in size. They existed; however, it is very difficult for the ladies and gentlemen of false science to have ever seen all of those reptiles from the Mesozoic Era. How did those dinosaurs live? How did those dinosaurs reproduce themselves? Were the ladies and gentlemen of false science witnesses of it?

The Cenozoic Era followed. It is stated that many reptiles evolved towards the state of mammals. It is asseverated that from the primates came the hominids, which are the ancestors of the intellectual animal mistakenly called a human being.

There is no doubt, state the ladies and gentlemen of false science, that the hominids were born of the primates, which gave origin to the human being, and also gave origin to the branch of great gorillas or chimpanzees, etc. When they talk like this, they practically use Darwin as their base.

However, we know very well that Darwin did not state in any way that the human being came from the ape. He only clarifies that the human being and the ape have a common ancestor.

Nevertheless, the materialistic anthropologists state that their common ancestors are the primates. They state that the first hominids were born of the primates as well as the great apes from ancient epochs, from the period that we call Cenozoic. This is how they arrange their theories so that these can coincide with Darwin's statements in some way.

Yet, did those primates exist by chance? Have the materialistic anthropologists ever seen, at any time, the first hominids? Can they swear that from the hominids were born the gigantic apes on one side and the human beings on the other? Can they affirm that from the primates the hominids were born and that these are the ancestors of the human being? Can they dare state that from those primates, all of the species of gigantic gorillas and chimpanzees which populated the face of the Earth in archaic epochs were also born? What do the ladies and gentlemen of materialism know about this?

On the other hand, as usual with his theories Haeckel assigned about seventeen or eighteen genealogies onto the present humanoids. He states that these gene-alogies come from marsupials or mam-mals. He states this theory with so much conviction as if he himself had indeed witnessed it. Haeckel and his partisans believe in phantasmagoric utopianisms of the mind that cannot be demonstrated.

The mammals, the marsupials, and all of those genealogies from Haeckel are extremely absurd. Amongst those genealogies, the anthropologists do not discard the famous lemurs with a discoid placenta. Yet, where is this placenta of the lemurs?

In these present times, they affirm that we come from the lemur; other certain pseudo-sapient anthropologists, who allege that our ancestor is the mouse, do not go unnoticed either.

Frankly, the materialistic scientists of this epoch have dedicated themselves to too much research. Their boasted materialistic culture from this century is indeed shameful; first they have us descending from the lemur, then all of a sudden from brother mouse.

Through ancient traditions we know very well that the human race was formed by the giants of Atlantis, of Lemuria, of the Hyperborean epoch, and of the Polar.

Yet, an excessive amount of ignorance has to exist in order to establish such theories that allege that we descended from the mouse, since in Atlantis the same mouse was not as small as it is in this day and age.

To state that the human being was tiny and that he kept on growing, that his stature was so small because he is a child of the mouse, is frightfully ludicrous in depth.

We can see how these cynical anthropologists from materialism act. One minute they are stating that we come from primates, then, all of a sudden they are stating another thing. When they get tired with brother mouse, they then appeal to the mandrill, based on the fact that this poor primate has red buttocks. Alas, these pseudo-anthropologists are so ignorant.

These pseudo-anthropologists are the ones who have precipitated this world towards the path of devolution and degeneration. They are the ones who are corrupting this humanity, the ones who are depriving people of their principles and the eternal values of their Spirit. When the eternal values of the Spirit are taken away from the human being, he then frightfully degenerates.

These pseudo-anthropologists are the ones who send their partisans (those scoundrels of materialism) to teach all of their nonsensical theories to the people of the villages. They are the ones who have converted themselves into the instructors of the wretched people who live in those villages.

We can consider these pseudo-anthropologists the corrupters of minorities, since with their display of fallacies they corrupt the simple individuals of the villages. They are the ones who harm the minds of people. They are the ones who structure educational plans, eliminating all of that which has the flavor of spirituality. Nonetheless, they do not have enough of a foundation to pronounce themselves against the mystical teachings of humanity.

It is abundantly ludicrous to asseverate in any given moment that we come from the mandrill or from the mouse or from the first primates of the Cenozoic Era.

The materialistic ladies and gentlemen from the North or from the South or from any of the different countries of the Earth laugh about the father Manu from whom the entire human race came forth. They laugh about the Dhyan Chohan who seems to be a utopian personage to them. Yet, they do not have any trouble believing in the fallacy of Haeckel. They believe in Haeckel's stupid pithecanthropoid with a speaking capacity that is a thousand times more fantastic than the assertion of Manu or Dhyan Chohan.

Despite the ignorance of materialistic anthropology, humanity still believes in the Dhyan Chohan, even when these illustrious materialistic anthropologists are upset about it. Still millions of people in the world accept the father Manu; people from Asia believe in him. The father Manu is the man-spirit; he is the prototype man who is placed in a very superior level of the Being.

Indeed, if we could perform retrospective investigations with procedures very much different from those of carbon-14 or potassium argon, then we would discover that truly, the prototypes of this humanity come from the superior dimensions of nature and the cosmos.

We have to judiciously analyze what this present materialistic culture that serves as a foundation for countries and nations)actually is. We must search for the origin of all this corruption and all this perversity. It is not possible for this humanity to continue being a victim of ignorance. To agree with their theoretical nonsense is one hundred percent absurd.

To state that those lemurs, those small animals with very enlivened eyes, had placentas and that they could be considered as one of our ancestors is an absurdity. It is an unforgivable zoological mistake, because lemurs never had placentas.

Indeed, Haeckel has harmed society a great deal. This is why we once stated (as a parody to Job the Prophet), "Let the memory of him be forgotten by humanity; neither let his name be uttered in the streets."

The science of embryogenesis was unknown in the epoch of Haeckel. This is why he dared to speak about lemurs with placentas; his statements were indeed incongruent.

When one investigates these things, one feels nothing more than repugnance for the school of materialism that is presently corrupting our culture. Schools of materialism are currently taking away the eternal values from this present culture; they are precipitating our society onto the path of perdition.

Which indeed is the ancestor of the human being from the Cenozoic Era?

Which are the ancestors of the paleolithic human being and which are his descendants? Did Darwin ever meet them? Did Haeckel or Huxley ever meet them? What does materialistic anthropology base itself upon in order to speak with so much authority about Homo Sapiens, the primordial man? To which epoch does Homo Sapiens belong?

Intending to find the fossil remnants of the primordial man, Huxley searched within the subterranean layers of the Quaternary Period; yet he searched in vain since these fossil remnants were never found. That is because the human being is more ancient than what the pigs of materialism suppose. The human being existed during the Cenozoic Period, as well as in the Mesozoic Period and in the Paleozoic Period.

The former statement will not be accepted by the materialists. They want the human being to come strictly from the Quaternary Period and in no way will they admit that the human being existed during the Cenozoic Era.

The moment has arrived for us to delve deeply and reflect, to have a maximum analysis. What do the pseudo-anthropologists know about life? What do they know about the processes of life during the Primary, Secondary, Tertiary, and Quaternary Periods?

Indeed, conventional anthropology is a building without a foundation. Just observe all human beings with their Monads and we will see how the whole little theater of Haeckel, Darwin, Huxley, Marx and their partisans would fall to the floor and be made dust.

I stated in my third chapter about the navel of the universe. What, our planet Earth has a navel? Why not?! When we are born, when we come into the world, we also come with a navel. As is the macrocosm, so also is the microcosm; as above, so below.

In the previous chapter I commented about the Omeyocan. But, what is the Omeyocan? The Omeyocan is nothing more than the navel of the universe. The Earth-Moon existed in a forgone cosmic day. The Moon had seas and mountains filled with life and vegetation. Moreover, it also had its Paleozoic, Mesozoic, and Cenozoic Eras.

All the worlds that are, have been, and shall be, are born, grow, get old, and die.

The Moon is a dead, cosmic corpse; this has been demonstrated by the astronauts who descended upon the lunar ground.

However, lunar life in itself, the living substance of matter, did not die; it continued processing itself in the fourth coordinate, in the fourth vertical, along with all of the seeds of all things existent. Later on that living substance processed itself within a fifth coordinate, then within a sixth and subsequently within a seventh. When that living substance immersed itself into the seventh, then it submerged itself within the bosom of the Abstract Absolute Space. That homogenous living substance, that Mulaprakriti of Eastern people, that primordial Earth, continued to exist. That living substance was a seed that could not be lost; therefore it was deposited within profound space. Thus, life continued to exist, yet in a dormant way within that seed.

That seed, the living substance, slept for seven eternities within the Chaos. To be more specific, this living substance slept for seven eternities within profound space. Much later, the electric whirl, the electric hurricane, the darkness and the wind (as it is stated by the people of

Anahuac) made that primeval world, that primordial Earth called Iliaster, inhabited. Then the *two* functioned with its opposites: positives and negatives, masculine and feminine. The Chaos emerged from that Iliaster; the sleeping living substance awoke. This is why it is stated that the hurricane, the tempests and the darkness prevailed within the Omeyocan, Yoalli-Ehecatl.

Yoalli-Ehecatl is the god of the wind, of the hurricanes, of the electric movement. Yoalli-Ehecatl is the macrocosm endowed within the microcosm. Everything is in incessant activity. This is how the Chaos is. It existed within the Omeyocan; the Chaos was the Omeyocan itself. Nonetheless, all of the possibilities will abide there until the Wholeness can make the Chaos fecund.

So when the Wholeness made the Chaos fecund, then the Limbus, that extraordinary Limbus, gave origin to all that is, was, and shall be. Since then, from the navel of the universe and throughout many dimensions, there were successive unfolded manifestations. Thus, evolving and devolving life passed through many dimensions; the merely germinal human being passed into the protoplasm and finally crystallized in the protoplasmic Earth.

The first manifestation was in the world of the mind, in the region of the cosmic mind, the universal intelligence. The second manifestation of all that is, has been, and shall ever be manifested itself in the second period. Then, as an outcome of the second, a third manifestation appeared in an ulterior dimension. Therefore, before life appeared in this physical world, it developed and involuted in three superior, extraordinary dimensions.

It is obvious that before life appeared upon our protoplasmic world, a creature very similar to any mammal or to any simian, yet indeed very different from apes, appeared from the species of animals. When the original or primordial human being achieved crystallization in a dense form, he then passed through a transformation in his morphology and appeared in the northern Polar Cap that in an aforementioned time was situated in the equatorial zone.

In my next chapter I will talk about the movements of the continents. We will

factually tell the world what Pangaea is.
Now, we will only limit ourselves to state
that before life became crystallized in the
physical world, life was already developed
in other dimensions. However, I want
to emphasize the following statement:
the real human being existed in the first,
second, and third root races, prior to the
existence of the intellectual animal mistakenly called a human being.

The intellectual animal, the one who
appeared in the Quaternary Period, is by
no means a human being, but an intellectual animal.

In my former chapter, I stated that
in Lemuria the real human beings lived
splendidly, yet some of them degenerated
at the end of their epoch; thus, they sexually mixed themselves with beasts. This
present humanity is the outcome of that
sexual mixture; the result is the intellectual animal.

This is the moment to understand
these very delicate matters, to understand that the human being is prior to
the Quaternary, Tertiary, Secondary, and
Primary Periods. Proof of this is that even
though all of the living species of archaic
times disappeared, the human being — or

better said, the humanoid – continues to exist.

The intellectual animal mistakenly called a human being was capable of surviving in spite of many storms, in spite of the revolution of the axis of the Earth, in spite of the events of Pangaea. If the animals, the reptiles, etc. from other Mesozoic Periods were capable of surviving, then this shows us that the human being is prior to all of these periods that have been designed by materialistic anthropologists.

We must profoundly and seriously reflect upon the Gnostic anthropological studies. Endow the intellectual animal with a Monad, which is that spiritual entity of which all the partisans of Darwin, Haeckel and Huxley have intended to deprive us, and we will see how their materialistic circus will definitely fall.

These are the times to unmask materialistic anthropology. It is the moment to restore the eternal spiritual values to humanity.

Chapter Five

There are facts, cosmic and geological events, that are worth studying in these treatises of Gnostic anthropology. There is no doubt that Gnostic scientific anthropology unveils all veils related with the origin of the human being and the universe.

Obviously, the mechanism of nature is portentous. However, we Gnostics will never accept the possibility of mathematics without a mathematician or a mechanism without mechanics.

Gnostics do not want to defend an anthropomorphic god according to the doctrine of the Judaic Jehovah, *"An eye for an eye and a tooth for a tooth."* We know very well that such dogmatism brings as a consequence or corollary and by opposition the materialistic reaction expressed by all type of atheists.

It is necessary to understand that any type of abuse is harmful for humanity. In ancient times people rendered worship to the gods, that is to say, to the intelligent principles of nature and the cosmos, to the Demiurge architect of the universe.

The Demiurge architect of the universe is not a human or divine individual. Rather, it is Multiple Perfect Unity, the Platonic Logos.

Unfortunately, in the august Rome of the Caesars and even in ancient Greece a type of religious degeneration occurred. When they abused worship of the gods, then as a reaction monotheism with an anthropomorphic god emerged. Later, that monotheism with its anthropomorphic god produced as a reaction modern materialism. Therefore, the abuse of polytheism brings as a consequence monotheistic anthropomorphism; that is, the belief in a Biblical anthropomorphic god. On the other hand, through time the abuse of monotheism originated materialistic atheism. These are the religious faces though which people always pass.

Frankly, in the name of truth, I consider that the moment has arrived in order to eliminate that monotheistic anthropomorphism that has originated very harmful consequences. If the religious clergies would not have abused that worship then modern materialistic atheism would not exist.

So, monotheistic worship emerged as a reaction. Unfortunately, materialistic atheism was also born as a reaction against monotheistic anthropomorphism. Likewise, the belief in an anthropomorphic god is the outcome of the abuse of polytheism.

Again, when the worship of the gods of the universe is abused, then by simple reaction the outcome is monotheism.

Nevertheless, we need to recognize the intelligent principles of nature and the cosmos. But, I repeat, we are not defending an anthropomorphic god.

It seems, by all means, that the recognition of intelligent principles can be shown through many scientific analyses. Let us observe, for instance, an anthill. There, we see the intelligent principles in complete activity. We see how those ants work, how they build their palaces, how they govern themselves, etc. The same happens with a beehive; their organization is astonishing.

Let us endow each one of the ants or each one of the bees with a Pythagorean **Monad** or with a Hindustani **Jiva**, and it is logical that, as a fact, the whole anthill or the whole beehive will make sense, because all of the creatures live through a **Monadic principle**. Thus, Haeckel, Darwin, and Huxley's materialism would become completely destroyed in the face of this concept.

We are not rendering cult to any anthropomorphic god. We only want to point out that intelligence must be recognized in nature. It is not absurd for us to recognize that nature is endowed with intelligence. The order that is there in the construction of a molecule or of an atom shows us the intelligent principles with complete, absolute clarity.

We are in the appropriate epoch in order to revise these principles. If we disagree with materialism it is because it does not resist an in-depth analysis. Obviously, materialism is total garbage.

The creation of the human being through mechanical processes is more incongruent than the belief of the creation of the Biblical Adam who instantaneously emerged from the clay of the earth; however, the first belief is as absurd as the second one. Both are incongruent.

We Gnostics recognize that there is intelligence in the whole mechanism of nature. There is intelligence in the movement of the atoms around their gravi-

The Pleiades as seen through a terrestrial telescope.

tational center, in the movement of the planets around their suns.

It is true, no lie, certain, that our Sun, the one that illuminates and gives life to us, is one of the suns of that great constellation that rotates around Alcyone. Since ancient times this constellation has been called the Pleiades.

There are seven suns that rotate around Alcyone; this is not strange. We live in a corner of the Pleiades, on a small planet that rotates around the Sun, a planet that is populated by intellectual animals, a very small world that we named Earth.

Generally, each of the suns from the Pleiades, each of those seven suns, gives life to the corresponding planets that rotate around it. It is true, we do not deny it, our planet Earth is a small world that rotates around the seventh sun of the Pleiades. It is no less true that the Pleiades need an intelligent directrix principle.

Naturally, the pigs of materialism only believe in the fat and in the lard. Materialists are eager to reduce the wretched three-centered or three-brained biped to a simple productive and consuming tridimensional machine. The materialists want to deprive humanity of its intelligent principles; by their own whim they want to deprive the whole human mentality of their eternal values, the values of their Being.

We comprehend perfectly that when humanity is deprived of the values of their Being, then it frightfully degenerates. This is what is happening in these moments of world crisis and bankruptcy of principles. The know-it-all ignoramuses of materialistic anthropology are obstinately precipitating the wretched people of this century onto the path of the most obvious perdition.

The Pleiades need a directrix principle; however, in order not to fall into anthropomorphism this time, it is better that we state what the directrix principles are, because anthropomorphism has been very fatal due to the fact that it produces materialistic atheism. Anyhow, the directrix principle of the Pleiades is plural, yet it has a representation that in no way would the pigs of materialism accept. I want to refer to the **equatorial astral sun** of the Pleiades. This sun is invisible through the lenses of our telescopes, yet it is visible for those who have developed the most extraordinary type of vision. This is the **intuitive Prajnaparamita** type of vision in its more elevated degree. The term *Prajnaparamita* is abundantly difficult, since it is Sanskrit and it is unacceptable for atheistic anthropology, yet it is real in its transcendence for true human beings.

The equatorial sun of the Pleiades intelligently coordinates all of the cosmic, human, plant, animal, and mineral labors and activities. It coordinates even this group of celestial bodies known as the Pleiades.

Indeed, the equatorial sun is the sum of intelligent principles that the partisans of materialism abhor. Yet, the world is a world and will always be...

Materialism always produces degeneration of the brain and the mind, devolution of the human principles, total decadency, and incapacity for the development of the objective reasoning of the Being.

The Pleiades with their sun constitute a beautiful panorama of the universe. The sun of the Pleiades is not a visible sun; it is the astral sun located in the fifth coordinate.

If we, the Gnostics, would accept but only three coordinates, if we were bottled up within the three dimensional geometry of Euclid, then we would be enemies of the Eternal One, just like the materialistic atheists who are like donkeys that only believe in the grass they see.

The intelligent principles of that astral sun maintain the Pleiades in perfect harmony. This is something that we do not ignore since we have methods and procedures for the development of certain transcendental faculties of the Being that allow us to see beyond the simple telescopes and allow us to penetrate beyond the microscope.

We must take into account not only the Pleiades, but the whole galaxy in which we exist. This great galaxy (the Milky Way) has a capital, which is the sun **Sirius**; a hundred billion suns, millions of worlds, moons, and meteorites are controlled by Sirius in an extraordinary way.

Unquestionably, the sun Sirius is gigantic. Close to Sirius there is a moon that is five thousand times denser than lead. That moon rotates around Sirius. Extraordinary radiations come from Sirius that affect cosmic matter. Yet, we must not deny that from that moon five thousand times denser than lead also emerge terribly infrahuman vibrations.

It can be stated that the radiations of Sirius affect all of the supra-heavens of any cosmic unit. On the other hand, the tenebrous infra-radiations of the satellite that surrounds it affect the infra-infernos that produce chaotic states in the mentality of human creatures. It engenders materialistic atheism, etc.

The galaxy in itself (the Milky Way) with all of its extraordinary organization, with its spiral shape, has the sun Sirius as its capital. Undoubtedly, the Milky Way needs the intelligent principles that govern it.

The **polar sun** comes into our memory right now. Obviously, in the polar sun are the intelligent principles that wisely control, govern, and coordinate this galaxy in which we live, move, and have our Being. It is related to a marvelous Spiritual Sun that completely directs the whole Milky Way.

Obviously, without the intelligent principles of the polar sun (even when the whole of this galaxy is controlled by its capital star Sirius and even when it would be intelligently governed) something would be lacking from it. It would lack the intelligence of the spiritual sun or of the polar sun that is the very foundation of all of those intelligent principles of the Milky Way.

Still, the whole matter does not end here; we have to go far beyond that. Einstein already stated, "The infinite has a limit." He also asseverated that the infinite is curved.

There is no doubt that there are many infinities. Beyond this infinite, there is another infinite, and much further beyond, there are many empty spaces between infinite and infinite. There is no limit to the many infinities. Our infinite, the infinite of Einstein, has around one hundred thousand galaxies; each one has approximately one hundred thousand suns, and millions of corresponding worlds. This is what is possible to perceive through telescopes. Yet, indeed, this infinite in which we live needs intelligent sovereign principles that coordinate it in order to avoid many types of collisions and failures, as much as possible.

Fortunately, the central sun exists, the sacred Absolute Sun. The intelligent directrix principles of this infinite are within this sacred Absolute Sun; within it, I repeat, we live, move, and have our Being.

Intelligence governs the whole cosmos, whether it is within the infinitely small or within the infinitely large. There is intelligence in the macrocosm and in the microcosm, in a system of worlds, in a beehive, or in an anthill. Cosmic intelligence abides precisely in each particle of this great creation.

We live here, as we have already stated, on a small planet from this infinite universe, in a minute world that rotates around the seventh sun of the Pleiades that has its mechanism governed by intelligent principles.

Unquestionably, the geologists who have studied a lot do not know about the living mechanism of this planet Earth. The belief that the continents on which we exist are fixed, firm, and immovable has always existed, but that concept is mistaken. We, the Gnostic scientists, know very well that the Earth is more similar to the structure of an egg than to a firm mass. If we open a bird's egg and observe it, we will then see that it has a yolk that is movable and that it sustains itself over the egg white. The same happens with the Earth; the continents are similar to the yolk. These continents sustain themselves over a clear, pasty, fluidic, and gelatinous substance. This continental yolk is not static, it periodically moves and rotates over it own axis.

The entire continents of America and Europe were once connected and now they are separated. This is what the materialistic anthropologists state about Pangaea. Yet, they ignore the rhythms, periodic movements, and true geological history.

There is an abundant amount of proof in order to demonstrate the movement of the continental masses. Once, Atlantis was in the ocean that bears its own name. The followers of materialism have placed a question mark about the existence of this continent. Nonetheless, Atlantis' existence has been duly demonstrated in a conclusive way by the true sages that from time to time have appeared upon the Earth.

The learned ignoramuses stubbornly assert that continent sank like a milk-layer sinks, yet that is absurd. Atlantis sank as a consequence of the revolutions of the axis of the Earth. Yet, the materialistic addicts do not know this.

The Atlantean catastrophe left our present continents in a very bad situation. Observe the continent of the Americas and you will see that from the side of the Pacific Ocean, the Americas incline as if they want to sink into the ocean; while on the eastern side they rise. This is similar to when a boat is sinking; it never sinks vertically, but always on one side.

The very Andes mountain range inclines towards the Pacific Ocean. Let us see Europe: there is no doubt that it wants to sink into the Mediterranean; it is more submerged towards its profundity. The same happens with Germany and Russia. The Asiatic continent is inclined as if it wants to sink into the Indian Ocean.

The continents were left damaged because of the great Atlantean catastrophe, which unbalanced the geological formation of our world.

We have spoken a lot about suns and catastrophes and about many such things. Yet, the suns of Anahuac (from the Aztec calendar) invite us to reflect because these are very interesting, namely the Sun of Fire, the Sun of Air, the Sun of Water, and the Sun of Earth. These suns mark terrible **cosmic catastrophes**.

It is stated that the Children of the First Sun, the Protoplasmic root race, perished when they were devoured by tigers. The tiger is a symbol of wisdom, this is clear!

It is stated that the Children of the Second Sun, the Hyperborean root race, perished when they were demolished by

The Aztec Calendar

strong hurricanes. This is in reference to that humanity who lived on that horse-shoe shape land that is around the North Pole.

It is affirmed that the Children of the Third Sun, the Lemurian root race, per-ished by the Sun of raining fire and great earthquakes.

The Children of the Fourth Sun, the Atlantean root race, perished by the waters.

The Children of the Fifth Sun, the Aryan root race, we, the people of this epoch, will perish by fire and earthquakes. This is how it will be, and this will be fulfilled shortly.

The Children of the Sixth Sun, the Koradhi root race who will inhabit the future Earth of tomorrow, will also perish.

After having spoken to you about the suns of Anahuac, we will now pass into a minor cycle. Naturally, there are always Primary, Secondary, Tertiary, and Quaternary ages.

We will not base these ages on the five root races that existed. This time we will base them on something distinct. We will base them precisely on the movements that the terrestrial yolk undergoes. We will base them on that geological movement that is processed periodically upon its own axis, on the movement of the continents over that pasty and gelatinous substance.

So, from the former points of view we can talk about the Primary, Secondary, Tertiary, and Quaternary ages. Likewise we can talk about the Eocene, a Primary Period unknown to people — Eocene, Miocene, and Pliocene.

We do not deny that among these types of catastrophes there have been other catastrophes also, like the terrible **glaciations**.

Atlantis marks the end of the Tertiary Epoch (at the end of Atlantis to be more specific). So, that Tertiary epoch was very beautiful because of its Edens; it was delectable because of its great paradises.

Many glaciations have occurred and there is no doubt that we are approaching another glaciation. There are catastrophes caused by the verticalization of the axis of the Earth, caused by the verticalization of the poles of the world. There are catastrophes that are the outcome of the movements of the continents, and then earthquakes occur, followed by the start of glaciations.

Five glaciations are mentioned that were processed in accordance with the movements of the continents. Yet, we must know that glaciations have also been produced because of the verticalization of the poles of the Earth. It is obvious that the catastrophes and glaciations are multiple.

If we were to state that the human being did not exist during the Miocene, Pliocene, and Eocene epochs, we would be asseverating something false. It is remarkable to see that while the archaic species of animals were extinguished, the human being nonetheless continued to exist. I am mentioning the word "human being" in a merely conventional sense, since we already know that the intellectual animal is not a true human being. Obviously, when speaking of people, we have to describe them in some way.

Terrible changes have occurred; yes, they have. Let us think about the human race that emerged from the Eocene, with their tropical weather. Likewise, let us think about that race that developed and unfolded during the Oligocene, with its mild temperature. Finally, let us think about that race who lived in the Miocene, with its cold weather, with temperatures so low they were approaching the last glaciations. What is fascinating is that in spite of so many glaciations and catastrophes, human beings continue to exist.

The Paleolithic human being still exists; incredible, yet it is true. All of the species of archaic animals disappeared, the enormous reptiles of the Mesozoic, and nonetheless, human beings continue to exist. How is this possible? How is it that all of the archaic creatures have died and the human beings are still alive?

The Primary, Secondary, and Tertiary epochs passed by and still we see human beings walking on the streets. This gives us enough authority in order to tell Darwin, Huxley, and Haeckel — who with their materialistic theories exercised too much damage upon humanity —that the human being existed a long time before the Paleolithic Era.

In the third chapter I commented about the navel of the universe, about the Omeyocan. I then compared it to the cosmic seed from which this planet was born. It is clear that before the Earth could physically exist, the Omeyocan developed itself in various dimensions. I want to state that within the Omeyocan, the navel of the world, the whole planet was gestated. This planet passed through various periods of activity in diverse dimensions before crystallizing in the present physical form.

The human being as a seed was developed from the Omeyocan and began crystallizing little by little through diverse dimensions until finally taking on physical form in the Polar Epoch.

Here we are stating subject matters that upset the materialists. They say they only believe in what they see. Nonetheless, they believe in all of their utopias. They are searching for the primordial human being within the subterranean layers of the Quaternary Period. Each day they invent more and more theories and they believe them without ever verifying them. They are going around uttering lies. They believe in what they do not see, therefore they are falsifiers.

We can prove what we are stating. We have systems of investigation by means of the technique of **meditation**. Through meditation we can develop certain faculties, such as Prajnaparamita, a type of intuition which allows us to study the Akashic Records of nature. The whole history of the Earth and of its races is found in those records.

If the pigs of materialism would leave their fanatical attitude and resolve themselves to enter into the Gnostic disciplines, then they would develop certain faculties which will grant them access to the history of the Earth and of its races.

The hour has arrived in which each one of us must reflect upon ourselves and the universe.

The human being existed upon the Earth far beyond the Primary Era, much prior to the Paleolithic Era. The concrete fact that we continue to exist in spite of the fact that the animals of forgone times have disappeared in their majority gives us the right to affirm the former statements. So if the former statements are a fact, then we have the right to state that we are as ancient as the Earth itself, as nature itself. Facts are facts and before the facts we have to surrender.

If we have not perished, if we have not disappeared from the scenario of the world in spite of so many catastrophes, if we have not disappeared even though the creatures from the Mesozoic times are extinct, then this gives us the authority to state that we are very special beings. We existed on the Earth before the creatures of the Pliocene or of the Mesozoic times ever appeared upon the face of the Earth. This right is given to us by the fact of existing, by the concrete fact that eternities have passed and we still continue to exist.

The contemporaneous creatures died and nonetheless we are still alive. All of them perished, yet we continue to exist. Therefore, we have a basis in order to laugh in the faces of Huxley, Darwin and Haeckel, personages who were lethal for humanity.

The different scenarios in which this humanity has developed deserve to be studied and to be kept in mind.

How marvelous and wise are the Nahuas' suns; they not only contemplate the root race who was devoured by the tigers of sapience, they also contemplate the Hyperboreans who were demolished by the strong hurricanes, the Lemurians who perished by the sun of raining fire and great earthquakes, and the Atlanteans who perished by the waters.

These Nahuas' suns go further; they contemplate those movements of the yolk over the egg-white; in other words, they contemplate the periodic movements of these continents that suddenly split or diverge, and then suddenly produce great glaciations where all life perishes, and then all of a sudden they originate new activities.

These Nahuas' suns work throughout the Tertiary, Secondary, and Primary periods. Finally, every period of fifty-two years, they are elevated in the changes of fire. We are now in the fifth of those changes, the fifth sun. The secret doctrine of Anahuac contains precious treasures that the enemies of the sacred land of Anahuac, the anthropologists of the atheistic materialism, would never accept.

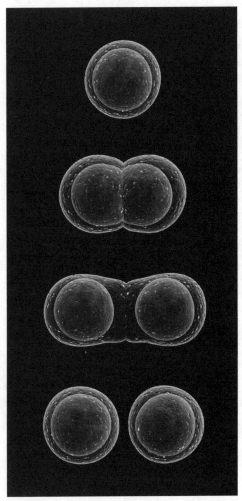

Cellular Division

Chapter Six

Unquestionably, this terrestrial humanity has passed through diverse phases of development; this is something that we have to judiciously analyze.

Intellectuals speak frequently about the mechanical evolution of nature, the human being, and the cosmos. Gnostics have comprehended that from the anthropological point of view there are two types of evolution.

Obviously, the first type of anthropological evolution, through alchemical sexual cooperation, would initiate its development when it is duly comprehended in each and every aspect.

The second type of anthropological evolution is different.

Unquestionably, in the beginning the human race multiplied themselves as cells multiply. We know very well that within the living cell its nucleus is divided in two. The living cell specializes in issuing a determined quantity of cytoplasm and inherent material in order to form new cells. The two, in their time, divide into another two. Thus by means of the fissiparous process of cellular division the cells multiply and thus the organism develops, etc.

In the beginning, the androgynous humans **divided themselves** in two in order to reproduce themselves. Later on that changed. The human organism had to become prepared (through evolution) in order for it to reproduce by means of **sexual cooperation**.

Obviously, it was in Lemuria (the continent situated in a foretime in the Indian Ocean) where the principal events related with reproduction through sexual cooperation were verified.

In the beginning, the creative organs, the lingam-yoni, were not completely developed. It was necessary for these organs of the human species to be totally crystallized and developed, so that later on in time the reproduction of the human species by means of sexual cooperation could be concretely performed.

Thus, while the masculine-feminine sexual organs became completely developed, many extremely interesting events from the biological and psychosomatic point of view were verified. These biological events did not occur in the merely androgynous humans but instead they occurred within the **hermaphroditic** human being.

The cell-atom was detached from the father-mother's organism in order to be developed and fully grown. Then as a consequence or corollary (by means of very delicate processes) a new creature developed.

The second aspect of this process of multiplication was remarkably peculiar. If it is true that in the beginning living human-seeds were detached as an atomic radiation in order to be externally developed and in order to become new creatures, then afterwards, in the second aspect, a certain favorable change was verified.

We know that noramlly each month the feminine sex eliminates an ovum from her ovaries; yet, in the hermaphroditic father-mother this ovum had a certain extraordinary consistency. It was already an egg in itself, in its intrinsic structure. It was an internally fecundated egg by the father-mother; it was an internally fecundated egg within the hermaphrodite. When the egg came out to the external world, it was developed or incubated until it finally opened and a new creature sprang forth from it. This new creature was nourished with the breasts of the father-mother; this is factually very important.

Much later in time, it was noticeable that certain creatures were coming into existence with one sexual organ more developed than the other. Finally, the moment arrived when humanity was divided into opposite sexes. When this occurred, sexual cooperation became necessary in order to create and to create anew.

The genealogies of Haeckel with respect to the possible origin of the human being and our three primordial races do not fit within materialistic anthropology which in this day and age is invading the world.

Regrettably, Haeckel's genealogies are the motive for laughter for conventional anthropologists who are enemies of the divine. Generally, they even mock the genealogy of Haeckel as well as other kinds of genealogies, such as that mentioned by Homer. Let us remember Achilles, the illustrious warrior, son of Mars, and Agamemnon, son of Jupiter, the one who commands from afar, etc. These are phrases or poetic words of that

Adam sees Eve after the Biblical symbol of the division of the sexes

man who in former times sang to ancient Troy and to the rage of Achilles the warrior.

We must speak clearly in this anthropological disquisition. It is obvious that the scientists of this epoch will have to decide if they are with Paracelsus, the father of chemistry, or if they follow the mythological folly of Haeckel. Anyhow, there is so much that we must acquire within the exclusively anthropological field.

If the division of the living cell as well as the primordial or primeval reproductive process would be denied, then as a fact the reproduction of Haeckel's Monera or atom from the aqueous abyss that divided itself asunder in order to be multiplied must also be denied.

Indeed, science cannot in any way pronounce itself against the primeval fissiparous act of reproduction, which is the reproductive system by means of cellular division.

Nevertheless, we are aware that these two theories about the way in which the reproduction of human species started are very arguable and thorny, namely reproduction of the human species before initiating the possible sexual cooperation when the creative organs were still in the process of development, and the reproduction of human species through sexual cooperation.

All the religious theogonies starting with the very ancient Orphic up until the Christian Bible's theogony, state a beginning by means of sexual cooperation, yet in a purely symbolic way. Those statements could be repeated by means of alchemy, but never in a scientific, anthropological way.

When the creative sexual organs have not yet been created, an evolving process cannot start through sexual cooperation. Obviously, in order to fecundate through sexual cooperation, there had to be a period of preparation. This had to have been a period through which the creative organs were developed and unfolded in the organic physiology of the human being.

The religious scriptures from the East as well as from the West have been adulterated, except the *Vishnu Purana*. For example, in this scripture it is stated that before human beings were endowed with the capacity of reproducing themselves through sexual cooperation or long before human beings could have that capacity, specifically before sexual cooperation between men and women existed, other ways of reproduction existed. This refers to former stages, previous to the formation of the creative organs in the human being.

We do not want to reach the point of affirming that those former systems previous to sexual cooperation did not have any relation with the sexual creative energy. We know that the sexual energy in itself has other forms of manifestation. Thus, before the creative organs were developed in the human species, the sexual energy had other ways of expression in order to create.

It is pitiful that the sacred scriptures of all religions have been adulterated. We understand that even the *Edda* generated a little alteration in the Pentateuch of the Hebraic Bible.

Anyhow, it is indispensable that we continue analyzing and meditating upon where the distinct races were developed.

We have stated many times before that the theory about the Pithecus-Noah is absurd, so is the cyanocephalus with a tail, the monkey without a tail, and the arboreal man. These are merely utopian matters that do not have any type of foundation.

We have already laughed about the silly theory of Haeckel that depicts a species of primate, something like a silly missing link between the ape and the human being, with the capacity of speaking.

However, it is necessary to know where these races were developed, in which scenario these evolutions and devolutions of humans unfolded. This is what we need to know because it would be impossible to disconnect the human races from their environment, from their distinct continents, from their islands, from their mountains, and from their natural scenarios.

It draws our attention to the fact that even though there were many varieties of animals during the Mesozoic Era and later became extinct, humanity is still very much alive. All of those animals have already disappeared from the face of the Earth; how is this possible? How is it that all the antediluvian monsters have disappeared, yet humanity continues to live on? We have made much emphasis on this statement, because it is indispensable to reflect a little about it.

It cannot be denied that the human being is related to his environment. It cannot be denied that there were other forms of reproduction distinct from those of sexual cooperation.

Therefore, it is convenient to know something about the environment in which all the different races were developed. It is urgent that we study the distinct scenarios of nature little by little.

We do not deny, in any way, that there are many facts that conventional anthropologists truly do not know. What do they know about the changes or modifications of the axis of the Earth in relation to the obliquity of the ecliptic?[9]

Laplace, the one who invented a famous theory that still exists in this day and age, affirmed that all worlds emerge from their correspondent nebulas, a theory that has never been proven. He even fanatically stated that the declination of the axis of the Earth in relation with the obliquity of the ecliptic is almost null. He stated that it has always been like that, always in a secular form.

Unquestionably, geology, up to a certain point, is against these former concepts of astronomy. Evidently, the deviation of the axis of the Earth within the obliquity of the ecliptic or to be more specific, the inclination, indicates glacial periods that always occur throughout the ages.

---

9   Obliquity of the ecliptic: "The angle between the plane of the ecliptic (or the plane of the earth's orbit) and the plane of the earth's equator; the "tilt" of the earth," represented by ε.

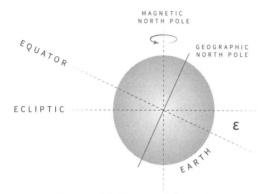

The axes of the Earth in relation to the
obliquity of the ecliptic (ε)

If we deny the glacial periods, then we
are affirming incoherent things, because
glaciations are completely proven. The
cause of these glaciations is the deviation
of the axis of the Earth on its inclination
within the obliquity of the ecliptic.

The former deviation, which astrono-
mers deny, is demonstrated with certainty
by means of geological studies. Proof of
tremendous glaciations exists. Already,
Magellan wrote about certain epochs of
heat or tropic within the Arctic, simultaneously accompanied by glaciations and
intense cold. We will conclude by stating
that, regarding this investigation, conven-

tional geology and astronomy are now in disagreement.

We have arrived to a very special subject-matter, the glaciations. It seems incredible that at other times in the south of Europe and in the north of Africa there have been terrible glaciations. In Spain, for instance, they know through accounts of the Silurian Period that there were gigantic glaciations there. This is demonstrated by means of the studies of paleontology.

No one can deny in this day and age that mummified cadavers of antediluvian animals have been discovered in Siberia and especially at the mouth of certain rivers like the Obi or Ub and others. This signifies that Siberia, which is so cold, was in other times a tropic with great heat; this is also true for Greenland, the peninsula of Scandinavia, Suez, Norway, Iceland, and all those horseshoe shaped lands that totally surround the North Pole. What? There was great heat in those regions? Impossible, everybody would say, yet paleontology has confirmed it.

Precisely, very interesting creatures have been discovered at the mouth of the river Obi. This invites us to reflect.

The northern regions of the Earth at this time.

During the times of Atlantis, the North and South Poles were not located where they are now. The North Pole (the Arctic) was located over the equatorial zone on the most eastern point of Africa and Antarctica.

The South Pole was exactly located over the same equatorial line, on the opposite side, in a specific place in the Pacific Ocean.

Therefore, there were great changes in the terrestrial globe's physiognomy. True ancient maps are known by the sages of this present epoch. Maps of the ancient Earth, geographic charts that demonstrate that our world had another physiognomy in the past, are in the secret crypts of the Lamas in the Himalayan mountains.

Let us think about Lemuria, about that gigantic continent located long ago in the Indian Ocean. It was united to Australia, because Australia was part of Lemuria, and likewise so was Oceania.

The Arctic was located in the most eastern point, over the equatorial line of Africa; then everything was different, distinct. A gigantic glaciation occurred during that epoch. That glaciation was projected precisely from the Arctic pole (located in Africa) towards Arabia, that is to say, towards the southeast of Asia. Moreover, it almost covered Lemuria completely; that entire zone was covered by ice, but it did not manage to pass into the Mediterranean.

It is intriguing to know that there are epochs in which our planet Earth passed through glaciations, epochs in which the ice invaded certain zones; epochs in which millions of creatures died. All of this, indeed, happens because of the inclination of the axis of the Earth in relation to the obliquity of the ecliptic.

The human being has had to develop himself in different scenarios, and we must deeply know what those scenarios are. How did America emerge? How did

Europe appear? How did Lemuria sink? How did Atlantis disappear?

Lemuria was accepted by Darwin. Lemuria still exists at the bottom of the Indian Ocean.

Obviously, the organisms have passed through different morphological changes, in these or those environments.

If we were to state (as the present materialistic anthropologists abundantly state) namely that the intellectual animal mistakenly called a human being has the famous mouse (or better said, the "runcho," or mouse which is cited by the South American people) as his ancestor, then frankly we would adulterate the truth. We know very well that such an enormous mouse or runcho from South America comes from the Atlantis of Plato. We also know that before Atlantis came into existence, the human being already existed. Therefore, the existence of the human being predates the famous Atlantean runcho.

If we would affirm that the human being originally came from certain primates and later on from certain hominids from the ancient land of Lemuria (which was so accepted by Darwin) then we would also adulterate reality. Because before the simians came into existence and a long time before the so-boasted primates and hominids appeared on the Earth, the human being already existed.

The human being already existed long before the reproduction of human species was developed through sexual cooperation.

So the human being is very much prior to Lemuria itself, before the Lemuria that was so accepted by Darwin.

We have to recognize that materialistic anthropologists have been studying this human race in a very superficial way, because this human race that has passed from Monolithic times through the Eocene, Miocene, and Pliocene epochs is more ancient than the continents of Lemuria and Atlantis.

Yet, it is necessary to continue studying the scenarios of our world in order to better comprehend the diverse processes of evolution and devolution of the different human races.

Meanwhile, I just want to state that the Gnostics are firm in their concepts. Therefore, if they have to choose between Paracelsus the father of modern chemistry and Haeckel as the famous creator of the silly mythic-anthropoid, frankly they will resolve themselves to the great sage Paracelsus.

The coelacanth, believed to have been extinct for 70 million years. Top: Fossil exhibit in the Houston Museum of Natural Science, Houston, Texas, USA. Bottom: Specimen caught the 18 October of 1974, next to Salimani/Selimani (Grande Comore, Comoros Islands), now preserved in the Natural History Museum, Vienna, Austria. Photo by Alberto Fernandez Fernandez.

Chapter Seven

The inhabitants of the planet Earth still do not know the world in which they live. However, incongruently, they want to travel to other worlds. Indeed, the planet Earth deserves to be profoundly studied.

Diverse facts, events, phenomena, completely unknown to conventional science are found on the entire planet. It is not irrelevant to consider some events that take place within the seas. Let us begin by describing some reminiscences about certain phenomena related to some of the creatures in the oceans.

In August 1917, a marine serpent that measured twenty-seven meters in length was observed off the coast of Cape Ann, Massachusetts (U.S.A.). The Naturalist Society of Boston observed it in detail; unfortunately, it was never again seen in those places.

While on a tugboat, the Oceanographer Danes Anton Brunn observed the capture of a tadpole (in a larval state) that measured two meters in length. According to naturalist studies, if this tadpole would have developed it would have reached twenty-two meters in length.

Anyhow, the above mentioned are two types of unknown creatures. Yet, how do they live? How do they behave? Where do they develop? Why do they exist?

Let us think about a famous blue fish that was considered to be extinct. This blue fish has always been referred to in a poetic way. Its peculiar name reminds us of a canto, of poetry; its name is "coelacanth." This strange little animal has extremities similar to those of human beings. It lives in the depths of the Indian Ocean; it existed in Lemuria. This indicates that even now, in this day and age, that fish continues to inhabit the same sunken Lemuria. It lives in the profundities of the ocean and very seldom does it surface.

Unquestionably, the great profundities of the Indian Ocean are extraordinary. The existence of an antediluvian animal during this century is something that makes us think quite a bit. Why does it still exist? What is the reason for it?

When the fossilized remnant of a coelacanth was found, scientists calculated it to be eighteen million years old. The coelacanth was very well known seventy million years ago. The figure of the coelacanth is astonishing; our attention is called to its residual extremities. Its limbs are developed similar to the arms, the hands or the feet of human beings. It still exists, and it is perfectly alive.

How is it that an antediluvian creature still lives in the final decades of the twentieth century? What could the materialistic anthropologists state about an animal such as this? What concepts would they have about it? All of this invites us to a great deal of reflection.

Indeed, what could be stated about the ichthyosaurs of archaic times? This animal still continues to exist in the terrible profundities of the Pacific Ocean. What do the scientists know about it? Scientists know absolutely nothing about it at all!

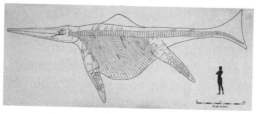

A common ichthyosaur fossil to scale with an average human being.

Thus, it is convenient for us to continue exploring all these subject-matters in order to attain clear concepts about them.

The case of the eels is certainly very special. Some eels from Europe and America meet in the Sargasso Sea with the purpose of reproducing. What is intriguing is the fact that the ones that return to that original point of departure are not the parents, but rather their offspring. Why does this phenomenon occur? Why do the parents not return, but instead their offspring? Can the anthropologists of false science explain this phenomenon? What do they know about this subject-matter? I am sure that they totally ignore these matters.

Let us study the case of the tuna fish. This is something that indeed deserves to be reflected upon. The tuna leave Brazil and swim toward Scotland; subsequently, they approach Europe and pass close to the Mediterranean Sea. Yet, it is rare for the tuna to swim into the Mediterranean Sea. What could scientists say about this phenomenon? Why don't the migratory currents of tuna enter into the Mediterranean Sea? Who guides them? Why do they do it? When have the gentlemen of materialistic anthropology defined these wonders, during which epoch? If they pretend to own the wisdom of the universe, then why have they not spoken about this matter? These materialistic people do not only ignore, but they ignore that they ignore, and this is very critical.

There are large squids, gigantic monsters endowed with enormous tentacles, but retarded science has never spoken about them.[10] The size of these squids can be calculated based on the skeletons that are occasionally found. Prints of their gigantic tentacles have been found on the backs of whales. Unquestionably, the squids' tentacles suck the pigmentation out of whales' skin and leave their print. Those prints reveal their titanic fights in the profundities of the ocean.

There are alligator fish (or ichthyosaurs of unknown origin) that pseudo-anthropology has never commented about.

Continuing ahead, we will also speak of certain phenomena that are unknown to those materialistic gentlemen. We know that deep down within the sea hundreds

10   Since the mid-2000's, scientists have begun looking at these creatures seriously. Before that, the scientific community mocked the notion of their existence.

of meters within its profundity, there are rivers that run in opposite directions. Why do they do it? Why does a river that runs close by another river circulate in an opposite direction and yet within the same ocean?

The rivers of the north circulate from left to right as do the needles of a clock; they circulate clockwise. The rivers of the south circulate inversely, counterclockwise. Why does the current of Bengal not rotate? What is happening? What explanation can pseudo-science give us? Why are the scientists speechless about these points? What can they tell us about this matter?

On the coasts of Peru (about 1,500 meters in profundity) engraved columns have been observed among Cyclopean buildings; magnificent photographs have been taken of them. Thus, the existence of Lemuria is demonstrated. Yet the foolish scientists continue as always, denying and denying.

In this day and age, there are civilizations that have disappeared, like the one found on Easter Island. There are monumental effigies, enormous human heads chiseled by titanic hands. However, materialistic science has never said anything; it is only speechless, dumb, silent...

What would we say about Antarctica? There is no doubt that there was a powerful civilization at the North and South Poles prior to the revolutions of axis of the Earth. Undoubtedly, within the ice of Antarctica, remnants of those very ancient cultures must still exist. The day will arrive when an archeologist's shovel will uncover them. Meanwhile present science does not give any explanations.

There are gigantic waves in tranquil and serene seas. These are isolated waves that have no reason for being. We are referring to those waves that are called "seiche." What is their origin? Are they caused by some underwater earthquake? How can these gentlemen, the materialistic scientists, explain this? What would the enemies of the Eternal One say about it? That there are waves in the boisterous seas, yes, we accept this; but that an isolated, gigantic and monstrous wave emerges all of a sudden in a tranquil sea without any known reason is something that has never had a scientific explanation. Nonetheless, such facts occur within the ocean and the materialists do not explain them.

Hundreds of thousands of earthquakes occur in the underwater mountain range at the center of the Atlantic Ocean where the continent of Atlantis formerly existed. Let us remember that terrible earthquakes and great seaquakes finished off the Atlantean continent. Even in this day and age, the trembling continues in the submerged Atlantis!

It is convenient to reflect upon all of these themes, since in their depth they are abundantly intriguing. Indeed, the materialistic anthropologists do not know the Earth. This planet continues to be a true enigma for them...

The spiked lobsters reunite in appreciable quantities in order to migrate; they then descend to the continental platform and then proceed towards the abysmal flatness to an unknown destiny. What do the presumed wise men say about this? What explanations do they give? Where do these lobsters go? What is their exact destination? Why do they perform these types of migrations? Enigmas that the learned ignoramuses do not understand!

The Earth was not always as it is now. Its physiological physiognomy has changed many times. If we examine W.

Scott Elliot's four maps, we will observe
that the Earth was completely different
one million years ago. Those four geographical maps deserve to be taken into
consideration. They are similar to the
four maps that are in some subterranean
crypts in Central Asia. Those maps are
unknown to the all-knowing ignoramuses
of materialistic science. These maps are
secretly kept with the purpose of preserv-
ing them intact, because the gentlemen
of conventional anthropology are always
willing to alter everything with the goal of
justifying their much boasted theories.

The first of these maps from W. Scott
Elliot calls our attention. It is very intriguing; on it we see how the world was eight
hundred thousand years B.C. Then, the
region of the brachycephalus of this present ultramodern precarious anthropology
did not exist.

On those W. Scott Elliot maps, we can
see that the Bering Strait did not exist as
it is today. The only thing that existed was
water that covered Siberia and Europe all
the way to France and Germany. These
lands were submerged in the depths of
the oceans.

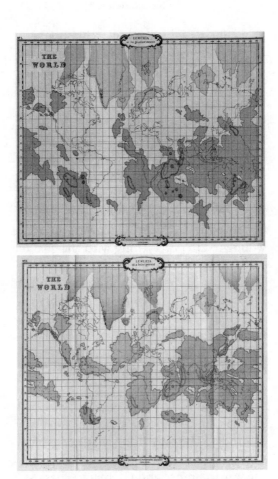

Maps of Lemuria by William Scott-Elliott
and The Theosophical Society (1904).

Africa, we can also observe, had only the eastern part because the west and the south were both submerged within the boisterous waves of the ocean. On those maps was a small continent in eastern Africa that was known at that time with the name of "Grabonski."

South America was submerged within the waters of the ocean. The United States, Canada, and Alaska were also submerged within the ocean. Nevertheless, on those maps the land of Mexico existed!

It seems incredible that eight hundred thousand years B.C. Mexico already existed. Europe still had not appeared and Mexico already existed! When South America had not yet emerged from within the depths of the ocean, Mexico already existed!

This invites us to comprehend that within the entrails of the sacred land of Anahuac, Mexico — which is as archaic as the world — there are archeological and esoteric treasures which have still not yet been discovered by archeologists.

In that epoch, Lemuria was a gigantic continent extending throughout the Pacific Ocean, covering the whole area of Australia, Oceania, and the Indian Ocean.

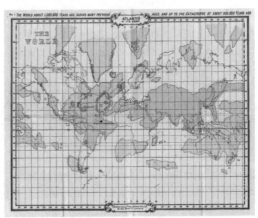

Map of Atlantis by William Scott-Elliott and
The Theosophical Society (1896).

Lemuria extended itself throughout the
Pacific Ocean reaching those coastal
places where much later South America
would emerge. Lemuria was monumental,
enormous.

Thus, the physiognomy of this terrestrial globe was completely different eight
hundred thousand years B. C.

The capital of Atlantis was Toyan, the
city of the seven doors of massive gold.

The materialistic anthropologists who
cannot see beyond their noses might ask,
on what do we base ourselves, speaking about the great capital of Atlantis? I
want to tell these gentlemen (who have

worked a great deal in order to take the
eternal values away from humanity, who
have precipitated it down into the devolving path) that there are maps preserved
in secrecy in subterranean crypts. These
maps indicate where Toyan, the capital
of Atlantis, was situated. Therefore, if we
talk we do it with complete knowledge.
If we locate Lemuria and Atlantis, it is
because they were continents that really
existed.

Toyan was situated on an angle, on the
southeastern part of that great country. It
faced the southeast coast of a strip of land
that visibly extended until the Loire in the
Mediterranean Sea and towards eastern
Africa. Its confines reached southern Asia
(which already existed).

Atlantis extended itself from Brazil to
the Azores and from Nova Scotia straight
down throughout the whole Atlantic
Ocean.

Thus, Atlantis totally covered the
ocean that bears its name. It was a great
continental country. Imagine Atlantis
extending itself from the Azores, up to
Nova Scotia and then descending towards
present day Brazil. How enormous that
continent was; it extended from south to

north; it was colossal! Yet, it sank through means of many earthquakes. Various catastrophes were necessary in order for Atlantis to definitively disappear.

The scenario of the world has changed very often; the physiognomy of this terrestrial globe has not always been the same. Different human races have been developed upon it.

We need to carefully study the physiognomy of the world of ancient times, the different geological changes through which the Earth has passed. Thus, only in this way can we form a precise concept about the origin of the human being and its diverse cultures; only in this way can we form a precise concept of its diverse evolving and devolving processes.

However, if we unfortunately remain completely bottled up within all of the contemporary prejudices, we will not be able to know anything about geology, much less about the development of the human race.

It is necessary to acquire, to investigate and to analyze... There are many enigmas on the face of the Earth that are unknown to conventional science. How is it possible that the ichthyosaurs that endured for

epochs (like during the Pliocene) continue
to exist in this present century in the
great profundities of the Pacific Ocean?
Until now these enigmas cannot be deci-
phered by the illustrious materialists, nor
can they comprehend them.

Through these lessons we must get
to know the different scenarios of the
world. We have to make light within
darkness. Thus, by stating the scientific
foundations of Gnostic anthropology
we will then be able to review all ancient
cultures. It is necessary to know how the
Pelasgians appeared in Europe. It is necessary to know about the archaic cultures.
It is urgent to know something about the
Hyperborean civilization, etc. Yet, first of
all, it is urgent to review the different geological changes that the Earth has passed
through.

We comprehend that each root race has
had its own scenario. We need to know
the environment, climate, and the conditions that each root race had in order to
be able to live. This is indispensable.

When materialistic anthropology states
that the American cultures came from the
Asiatic continent by means of the Bering
Strait, indeed they are asserting a frightful

mendacity, because ancient maps show that in ancient times the Bering Strait, Siberia, Canada, and the United States did not exist!

Mexico had a solemn, marvelous population that was separated from the Bering Strait by the great oceans of that time. Therefore, materialistic science is talking about something that they have not seen, that they cannot prove.

We are stating these facts based on maps such as those of W. Scott Elliot and other similar maps which are found in the subterranean crypts of the Himalayan mountain range in Central Asia.

People who asseverate that the human race reached America through the Bering Strait are showing a great deal of ignorance, a total unknowingness of ancient geographic charts.

Materialistic anthropologists are cheating the public with their types of assertions; this is how they insult the intelligence of readers.

We repeat: we love scientific investigation and exact analysis. We do not allow ourselves the luxury of accepting materialistic theories. We are not witless people who would allow ourselves to be cheated by superstitions based on false utopias.

We have geographic charts and we are sure that the readers of this book will comprehend our position very well.

The readers will comprehend all of this even better when they study the aforementioned geographic charts, which were displayed at the Third Gnostic International Congress in Caracas, Venezuela. That congress took place August 11-19 in the year 1978.

Anthropogenesis
A Lecture by a Gnostic Instructor

The human root races and subraces of our planet Earth are related with the law of Heptaparaparshinokh, the law of seven, which is the law that organizes. Every Kosmos in the universe is organized by the intelligence of the Logos; as, in the involution of the spirit into matter, through the mahamanvantara (the great cosmic day), within which occur seven manvantaras (cosmic days) . The cosmic days are also called rounds; this is why sometimes we address them as cosmic rounds.

Each cosmic day then has within it seven root races, and each root race has within it seven subraces. As an example, at this moment we are in the seventh sub-race of the fifth root race of the fourth round.

The present mahamanvantara or great cosmic day is the fourth, and is called the Terrestrial Epoch. Previously, there were three cosmic days, which Kabbalistically were related to Saturn, the Sun, and the Moon. Samael Aun Weor explained in detail these rounds or epochs in his book *The Revolution of Beelzebub*.

· First cosmic round: Saturn

· Second cosmic round: Solar (Sun)

· Third cosmic round: Selene (Moon)

· Fourth cosmic round:
  Earth (Terrestrial)

Every round recapitulates previous rounds before it then expresses its own character. So, we will say, in order for any cosmic round to express its nature, it has to recapitulate the previous rounds. In a similar way, a woman is a "recaptiluation" of her parents, grandparents, great grandparents — of course, the way she recapitulates their influences is on many levels: in her appearance, culture, mannerisms, genetic predispositions, inheritances (biologically, externally), etc. Obviously there is a great deal of fluctuation and change, nevertheless, no one can deny that our individual life is founded on the basis provided by our ancestors, mostly on levels we cannot even perceive. So, this also occurs on a cosmic scale in the unfolding of worlds, and also happens on the scale of a civilization. This is how we understand this in Gnosticism. The human being is a Microcosm of the Macrocosm. So, when the human being emerges from creation, it reflects the previous manifestations of the human being — this is true individual level, and also on the level of the human race as a whole.

The book of Genesis in the Bible was written by Moses, where he kabbalistically and alchemically described in detail the origin of this present Terrestrial Epoch. Moses also showed the way we can develop our human nature in order to become a Microcosmos, or what in Kabbalah is called the Zauir Anpin, "the lesser countenance" of the Arick Anpin, "the great countenance."

This Terrestrial Epoch was synthesized by Moses in the first verses of the Bible, when he says,

> *"In the beginning Elohim created the Heavens and the Earth."*

Elohim means "gods and goddesses."
On the Tree of Life of Kabbalah, we find the name Elohim applies to many aspects of divinity. Elohim Sabbaoth is the name of divinity related to the sephirah Netzach. Netzach is related with the

mind and the "mental plane," an aspect of the fifth dimension.

Netzach is directly related with the sephirah Binah, ruled by Saturn. Thus, when we study the sephirah Netzach, we find the origin or the beginning of this mahamanvantara: in the mental plane. As Samael Aun Weor explained in the book *The Revolution of Beelzebub,* the mental plane was where the root races of the Saturnian Round developed, in what they called "Arcadia."

The Saturnian cosmic day is described in Genesis in the very beginning where Moses says, "In the beginning Elohim..." Here, the word Elohim refers to those Monads that were self-realized during that epoch.

It continues, "...created the Heaven and the Earth. And the Earth was without form and empty." This describes the "protoplasmic state" of the Earth at that time.

Then, the book of Genesis states, "And the spirit of the Elohim, or the Ruach Elohim, was hovering above the face of the waters." That water is the pure Akasha, the primordial force, in the superior dimensions.

There are worlds, planets, in a protoplasmic state. Thus, in ancient times our
planet was in a protoplasmic state; firstly,
during the cosmic scenario called the
Saturnian round, and then several cosmic days later when the Terrestrial round
began by recapitulating that primordial
protoplasmic state. Both eras were inhabited by protoplasmic human beings, yet
each at their own level of evolution.

When the present Terrestrial epoch
began, the first race to emerge was the
Protoplasmic root race, which existed
within the superior dimensions of nature,
in a type of a protoplasmic matter that is
not physical. That root race was just the
recapitulation of the Saturnian epoch.

The Polar Root Race

The Protoplasmic human beings were
humans in the complete sense of the
word. Human comes from Hum, a word
that means Spirit, and Manas (man)
is mind in Sanskrit. The Protoplasmic
humanity was made of individuals whose
mind was controlled by their own particular Monad, Spirit. In other words,
in complete Sanskrit terms, they were

Bodhisattvas — that is, human vehicles of superior Beings, who became masters in previous cosmic days. Therefore, in that epoch there were no intellectual animals, only real human beings.

The protoplasmic type of matter that formed their bodies was elastic, ductile. They had the power to become small, to shrink themselves to the size of an atom, or to enlarge themselves to gigantic sizes, to even touch the Moon, things that are impossible for us. Of course, they did not exist in this tridimensional world, but in the superior dimensions of nature.

Remember that verse, "And the Spirit of God was hovering upon the face of the waters." If we associate that verse with our present humanoid level, we will say that this addresses the protoplasmic state of the fertilized egg, a stage at which we ourselves recapitulate the past.

Even now, we still have protoplasmic bodies in the internal worlds — when you dream you use that body, and it also can stretch, fly, shrink or expand, etc. However, our protoplasmic bodies are not of the same nature as those from the first root race; they had solar bodies, the bodies created through alchemy, called by

Jesus "the Wedding Garment." The protoplasmic bodies that we have internally are
lunar, animal, inferior.

In our sexual glands we also find the
Spirit of the Elohim that hovers above the
sexual waters. Above in the Macrocosmos,
those waters are the Akash; below, here
in our physical body, the Microcosmos,
we find the Akasha in our sexual matter where the energy of the Spirit of the
Elohim floats in us. If we preserve and
transmute those waters, we then create
internally.

The Hyperborean Root Race

The book of Genesis continues,

> *"And Elohim said, Let there be Light, and
> there was Light."*

That light is referring to the second
round (the Solar Round) and also to its
recapitulation in the current Terrestrial
round as the second root race, which
appeared in a lower dimension, yet still
within the superior internal dimensions
of the planet Earth.

The second root race was known as
the Hyperboreans, the people "beyond

the north wind." The Hyperborean root race was the recapitulation of the Solar Round; most of the beings who incarnated in that epoch were archangels. The archangels were the outcome of the Solar Round or Solar Epoch, or the second manvantara of this great mahamanvantara.

Both the first and second root races were androgynous beings whose bodies contained the two sexual forces. At that time, the sexes were not yet separated.

Androgynous comes from the Greek androgynos, "male and female in one," from andros "male" + genika "female", meaning a creature having male and female sexual polarities or characteristics. An androgynous being is a "sexless" being capable of reproducing its own species by means of asexual reproduction, in other words, a living species that without having evident sexual organs is capable of reproducing its own species (i.e. fissiparous "Tending to break up into parts or break away from an androgynous body").

No remnants of those root races can be found in the physical world, since they existed in the fourth and fifth dimensions.

Those root races are what we call in Kabbalah "Adam Kadmon," beings that were not tridimensional, yet existed. They built their civilizations, and had contact with humanities of other planets. We have to understand that humanities exist not only in the tridimensional world, but also in the superior worlds; civilizations exist not only in the tridimensional world but in the superior worlds as well. The technology of such civilizations allows them to travel through space and time.

The Lemurian Root Race

The Lemurian root race was the first race to appear physically. Their bodies were semi-ethereal and semi-physical, between the third and the fourth dimensions, because they were descending from the fourth dimension into the third dimension, little by little.

The Bible called this the terrestrial paradise, where Jehovah Elohim placed the human being.

When you find in the Bible that Elohim said, "Let **us** make man in **our** image, after **our** likeness," [Gen 1:26] that is how the Bible explains how the Lemurian root race, created in the image of the Elohim, crystallized, condensed, descended, little by little into the physical world, that terrestrial paradise called "Eden." That was the primordial land that appeared physically on the planet Earth, crystallized from the Akashic waters of space.

That humanity of Eden was a gigantic, hermaphroditic root race, which the Bible called the Nephilim, "giants." They were the Lemurians.

We refer to the Lemurians not as androgynous but as hermaphrodites, which means each individual had the feminine and masculine sexual organs. There were millions of Lemurians, Nephilim, giant hermaphrodites, whose bodies contained both sexual organs.

So, the Lemurians were what the Bible in Genesis 5:1, 2 addresses as Adam:

> "This is the book of the generations of Adam. In the day that Elohim created Adam, in the likeness of Elohim made he him; Male and female (hermaphrodite) created he them; and blessed them, and called their name Adam, in the day when they were (physically) created."

And cosmically speaking:

"And Jehovah Elohim formed Adam of the dust of the ground."

The dust that the Bible is referring to is the plasma, protoplasmic elements that were crystallizing from the superior dimensions.

Next the Bible explains the division of the sexes that occurred after the fourth subrace of the Lemurian epoch. At that time, humanity was guided by the Elohim, they were in contact with the Elohim. Samael Aun Weor stated that the Lemurians were not different from the animals in the sense that internally they were in contact with their own particular Monad. The difference between the Lemurians and the animals was that the Lemurians did not fornicate [orgasm]. When the Lemurians were separated into sexes, their inner Elohim, and the archangels of that epoch, those who we call Kumaras, or divine kings of nature, were guiding them and teaching them about sexual cooperation.

When the Lemurian race was divided into Adam and Eve, or as we say Adam Chavah, the sexual act was necessary in order to maintain the multiplication of that race. Before that, when they were hermaphrodites; the sexual act was not necessary in order to multiply the species. The man-woman, or the hermaphrodite being, fecundated itself from its own testicles; the sperm was fecundating the ovum in its own ovaries. They were bringing children into the world without pain; the father-mother gave birth, and the hermaphrodite baby fed itself from the breasts of the hermaphrodite father-mother. That is how the Lemurians multiplied in the beginning.

When the sexes were divided, the male (Adam) retained the sperm and the female (Chavah or Eve) retained the ovum. In order to multiply they needed to make sexual contact to extract one sperm from the sexual masculine glands in order to place it in the ovum of the feminine sexual organs. Since they were new to the sexual knowledge, they were instructed by the Elohim.

The Elohim were not only physical, but were internal. Lemurians were able to see them. It is not like us today, we can see physically but not internally; we need to restore our internal senses to see the inner Being of any creature. At that time, the Lemurians had all twelve senses working;

the seven chakras were fully developed and through these senses they were not only seeing the Elohim of this planet Earth, but they were capable of seeing the Elohim of other planets, the humanity of other planets. These were senses that for us are incredible, yet for them were normal.

So, via their senses, physically and internally, they were guided by the Elohim. They knew how to perform the sexual act under the command of these Elohim, to which the Bible refers as Jehovah Elohim. This Jehovah Elohim is not one, but many entities, master Monads that were guiding humanity.

Under the guidance of the Elohim, the sexual act was performed in the temples, never in their homes. At that time humanity did not have what we call lust, cupidity, or sexual degeneration. They performed the sexual act only under the guidance of the Elohim, performing what we call in this day and age White Tantra. Men and women were united in the temples and they did not reach the orgasm. The Lemurians knew that only the beasts of the animal kingdom reached the spasm or orgasm, because animals are guided by instinct, they have no intellect, no reasoning to comprehend that fornication is a fault, something that they should not do. However, since the Lemurians had objective reasoning, they knew how to perform the sexual act without orgasm. Their children were born without pain. The Lemurians were not slaves of sexual desire. The Lemurians did not have desire at that time.

But, one day, infected by the Luciferian vibrations of nature as well by certain demonic elements, they were tempted to perform the sexual act out of the temples and without the guidance of the Elohim. They were tempted by "the serpent." That tempting serpent was not outside of them, but inside; it is called the Luciferian vibration. This inevitably led them to the reaching of the spasm, orgasm, because at that time they could not control that serpent that was circulating through their physical bodies. The lure of the serpent was within, just as we today feel it.

The outcome of the orgasm was the loss of the spiritual powers, as Jehovah Elohim said:

> *"My spirit shall not always strive with Adam, for that he also is flesh."* —Genesis 6:3

In other words, Elohim knew that sooner or later, humanity was going to fall into the sin of orgasm, or the sexual animal spasm. This is the why we learn in White Tantra that we have to perform the sexual act without the spasm, without the orgasm.

White Tantra is the method in which the couple controls the sexual force in order to develop objective reasoning, in order to develop their brain, and to acquire the power of controlling the sexual seed, in order to bring creatures into the world without the crime of reaching the orgasm of animals.

Thus, that humanity, who didn't possess much knowledge of that which the Bible calls "Good and Evil," were tempted by that energy of Daath in the Tree of Life, named the Tree of the Knowledge of Good and Evil, which is a reference to the sexual organs. Note how the Bible uses the words "knowledge, know, knew" as a reference to sex.

They were also tempted by the ability to have children on their own; regrettably, the males could not take one single sperm, as they had done before through the power of the Holy Spirit, instead they discharged millions in order to engender a creature, which is precisely the sin against the Holy Spirit. This is what is called fornication.

As we have stated, at the animal level, the animals spill the semen because they are instinctual, they do not reason, they do not have intellect, they do not know what is right and what is wrong, but we, animals with reasoning, should know what is right and what is wrong. So, in order for a male to take one sperm from his sexual glands and to place it into a woman's uterus to engender a child, we need a power superior to the mind, and that power is called in Sanskrit Kundalini, or what the Bible calls the power of the Holy Spirit. That Holy Spirit or Kundalini is a sexual power that the Lemurian humanity had been able to control. So, their interior Elohim were capable of taking one sperm from the physical organ of the man and engender the female during the sexual act without reaching the orgasm; that is a power. The female still retained the capacity of releasing one ovum every month or every other month. That power now is instinctual in women, but at that time in Lemuria, it was a con-

trolled power; the woman had the power
to release one ovum and the man had the
power to release one sperm. That power is
called the power of the Holy Spirit, which
is based in the spinal column.

The spinal column is where the Holy
Spirit stands and works with the seven
senses of the soul. So, when the Holy
Spirit as energy is active in the spine
from the coccyx to the brain, then all
seven chakras are in activity, and through
those chakras the Holy Spirit controls the
physical body, and the human soul has
the power of releasing one sperm without
the orgasm. But, when at that time they
committed the crime of eating from the
fruit of the Tree of Knowledge, which is
sexuality, they lost that power, because
the power that sustained that fire in the
spinal column is the sexual Akasha, in the
sexual organs. When that sexual semen,
Akasha, is ejaculated in the bestial way,
when millions of sperms have been ejaculated, the fire descends, because there is
no more sustenance to sustain that fire in
the spinal column. That is why it is written that Jehovah Elohim said to the couples that when you reach the orgasm, the
spasm of the animals, or, when you eat

from the fruit of the tree of knowledge, you will die. This means you will die in the Spirit, and you will be born as a beast, and this is what happened in that epoch of Lemuria. By eating from the forbidden fruit, by eating the fruit as the animals do, they lost the fire, and by losing the fire in the spinal column (the Kundalini, the Holy Ghost), they lost their inner senses, and the power of creating without the orgasm. The outcome of this was they began to reproduce like the animals.

So, since that epoch, humanity has created like the animals. If you observe a dog, a horse, a bull, a donkey, they fornicate in order to multiply. If you observe the intellectual animal of this epoch, they do the same in order to multiply. There is no difference. Perhaps, the only difference might be that animals do it once in a while, when they are with the lure of the sexual act, when they are sexually excited, which is every other week, every other season, while we do it constantly and every day just for fun.

So, that was precisely the crime of that humanity, and we say that humanity because they were really humans. As it is written symbolically, in other verses of the Bible:

> *"And it came to pass, when Adam began to multiply on the face of the earth, and daughters were born unto them, that the Beni Elohim (children of the Gods) saw the daughters of Adam that they were fair; and they took them wives of all which they chose.*

> *"There were Nephilim (Lemurian giants) in the earth in those (Lemurian) days; and also after that, when the Beni Elohim (children of the Gods) came in unto the daughters of Adam, and they bare children to them, the same became Giborim (mighty ones) which were of old age, humans of renown."* —Genesis 6: 1, 2, 4

When the children of the Elohim saw that the women were beautiful, they took wives from those daughters of Adam, and they fornicated with them and created children.

Beni Elohim is Hebrew for "the children of the Gods," a reference to the Bodhisattvas, those that were vehicles of great masters of the past.

In the book of Enoch, it is written, symbolically, about the 200 Beni Elohim, or children of the Gods that fornicated at that time and fell:

> "It happened after the sons of Adam had multiplied in those days that daughters were born to them, elegant and beautiful. And when the angels, the sons of heaven, beheld them, they became enamored of them, saying to each other, Come, let us select for ourselves wives from the progeny of Adam, and let us beget children.

> "Then their leader [the angel] Samyaza said to them; I fear that you may perhaps be indisposed to the performance of this enterprise; and that I alone shall suffer for so grievous a crime.

> "But they answered him and said; We all swear; and bind ourselves by mutual execrations, that we will not change our intention, but execute our projected undertaking.

> "Then they swore all together, and all bound themselves by mutual execrations. Their whole number was two hundred, who descended upon Ardis, which is the top of mount Armon. That mountain therefore was called Armon (curse), because they had sworn upon it, and bound themselves by mutual execrations.

"These are the names of their chiefs: Samy-aza, who was their leader, Urakabarameel, Akibeel, Tamiel, Ramuel, Daniel, Az-keel, Saraknyal, Asael, Armers, Batraal, Anane, Zavebe, Samsaveel, Ertael, Turel, Yomyael, Arazyal. These were the prefects of the two hundred angels, and the remain-ders were all with them. Then they took wives, each choosing for himself; whom they began to approach, and with whom they cohabited; teaching them [in Yesod] sorcery, incantations, and the dividing of roots and trees. And the women conceiving brought forth giants, whose stature was each three hundred cubits. These devoured all which the labor of Adam produced; un-til it became impossible to feed them; when they turned themselves against Adam, in order to devour them; and began to injure birds, beasts, reptiles, and fishes, to eat their flesh one after another, and to drink their blood (fire). Then the earth reproved the unrighteous." —Book of Enoch 7

Thus, that was the outcome when the Lemurian race fell into animal generation. Before its downfall, that humanity multi-plied through human generation, which is White Tantra. With the fall of the Lemurian humanity into animal generation, the black magicians, the Luciferians from other epochs, started teaching Black Tantra, or the way in which the human being acquires abysmal powers through fornication, without the power of the Holy Spirit but its opposite. That was the great mistake of that epoch. It is stated in the book of Genesis that at that time humanity still knew and practiced white Tantra:

> "And Adam **knew** Eve his wife; and she conceived, and bare Cain, and said, I have gotten a man from יהוה (Iod-Havah). And she again bare his brother Abel. And Abel was a keeper of sheep, but Cain was a tiller of the ground." – Genesis 4:1

In Hebrew, Cain (Kayin) means smith, which relates directly to alchemy. The alchemists of Lemuria, under the guidance of the Elohim, were developing objective reasoning. Cain, the mind, or solar mind, was created in the positive way, and through alchemy it is stated that Adam and Eve had another son, Habel, or Abel, the human soul was developed through White Tantra. But, in the process of time, when that humanity fell into fornication, their minds became polluted. That mind is Cain. The Bible continues,

> *"And Cain talked with Abel his brother: and it came to pass, when they were in the field, that Cain rose up against Abel his brother, and slew him." – Genesis 4: 8*

This statement, written in the Genesis 4:8, should be interpreted as follows: "And Cain rose against Abel *in the field of Yesod"* because it is stated that when they were in the field that "Cain rose up against Abel his brother and killed him." The field which the Bible is addressing is a reference to the sephirah Yesod, the sexual act. This means that in the moment when that humanity fornicated, their own particular individual Cain, their mind, killed their own soul. This is how we have to comprehend the mystery of Adam and Eve, Cain and Abel.

Adam and Eve, as explained in *Sex: The Secret Gate to Eden,* has many symbolic meanings:

- First, they are a reference to the Lemurian root race

- Second, they relate to Ida and Pingala in the physical body

- Third, they are also a reference to the brain and genitalia

When we talk about Adam and Eve, we have to understand them in that triplicity. When we read that Adam and Eve had their first son, Cain, we have to understand that is a reference to the mind: the outcome of alchemy. But, that mind, Cain, that in the beginning had been the outcome of White Tantra, became black with the sexual ejaculation; thus, during the moment of ejaculation is the moment when Cain (the mind) turns against Abel, the human soul, and kills him; this occurs inside the psyche. Thus, Genesis 4:8: "when they were in the field, that Cain rose up against Abel his brother, and slew him" is a psychological statement. Yet, it is also a general statement, a global statement; for Cain and Abel are the outcome of that hermaphroditic division, which is a division of that Lemurian race into Black Tantra and White Tantra. This is why we refer Cain and Abel as the separation of the psyche in two tantric ways, namely:

"And יהוה (Iod-Havah) had respect unto Abel and to his offering: But unto Cain *and to his offering he had not respect."*

- Genesis 4: 4, 5

This alchemical statement also happened at the end of the fourth subrace and the beginning of the fifth subrace of Lemuria.

Do not fall into the mistake of thinking that Adam and Eve was one couple that had two children; that is just a very literal interpretation of that alchemical book of Genesis, which is symbolic. When you read about Cain and Abel, you have to look inside, to your psyche: Cain, the mind. Abel, the soul. That is how you have to interpret them.

The soul (Abel) dies, because it is sunk into the Klipothic levels of nature (Hell). The Soul dies to the Spirit, and it is born in Hell. Cain (the mind) takes over, but it is a fallen Cain. All of this process took place during the Lemurian Epoch.

At that time, the Elohim tried to resolve the problem. The Elohim began to organize groups in order to guide the Lemurians and to lift them out of animal generation. This is written in the book of Genesis:

> *"And Adam* **knew** *his wife again; and she bare a son, and called his name Seth: For Elohim, [said she], has given me Seth as another seed instead of Abel, whom Cain slew. And to Seth, to him also there was born a son; and he called his name Enosh: then began men to call upon the name of* יהוה *(Iod-Havah)."* – Genesis 4:25, 26

Seth is the outcome of the repented fallen Lemurian race, who after knowing their problems, rose again through White Tantra (Daath, "knowledge"), and the outcome was another seed, Seth, which is the third aspect. Yet, not all of the Lemurians entered into White Tantra; remember that Cain, after the fall, also found his wife, and multiplied.

> *"And Cain* **knew** *his wife; and she conceived, and bare Enoch."* - Genesis 4:17

Likewise, when you read the fifth chapter of the book of Genesis you find that Seth, the third soon of Adam and Eve, the Lemurian root race through time, also had another son, who is named Enoch.

> *"And Enoch walked with God: and he [was] not; for God took him."* – Genesis 5:24

When translated, the word Enoch is a reference to one who is illuminated,

educated, or a seer; this means that when
the Lemurians were returning to White
Tantra they gained knowledge of good
and evil, because they were fallen. When
they rose again they were developing that
which is called "Seth."

Seth is related with the Kundalini
rising in the spinal column, and also
with the knowledge of good and evil. In
Egyptian mythology, Seth is a symbol of
that energy that rises.

Seth has a variety symbolic meanings
that are all connected; it is related with
the ego, the "red devils of Seth"; it is also
the symbolism of the fire rising in the spinal column, giving birth to human entities. Seth also represents those individuals
who were the seeds for the next root race.

The Atlantean Root Race

Seth-Enosh is the origin of the human
beings of the Golden Age of the Atlantean
civilization, yet Atlantis had two divisions, because Cain also had a son, whose
name was Enoch. Sometimes, Kabbalists
mislead the reader and instead of Enoch
they write Enosh, but both have the same

meaning. The Greeks call him Enoichion, which means the "seer of the open eye."

The meaning of Cain's son Enoch is that he was also a seer — not a seer of the divine but a seer of Klipoth, hell, which is the type of seer that we find everywhere in this day and age.

There are two ways of seeing. When you awaken your consciousness in a positive way, you start to see in all the levels of the Tree of Life, but when you awaken your consciousness through Black Tantra, or drugs like LSD, cocaine, heroin, mushrooms, marijuana, and all those drugs that are common in this day and age, you also become a seer, you also see the other dimensions, but not the higher dimensions; instead, you see the Klipothic, inferior, infernal dimensions - hell, in other words. So, there are two ways to awaken: up and down, and this is the difference between the Enoch of Seth and the Enoch of Cain.

It is written that Seth was "another seed instead of Abel, whom Cain slew." This means that Seth is that consciousness (Abel) that is rising in the psychology of those Lemurians who practiced White

Tantra, a consciousness with more knowledge of good and evil.

Regrettably, this is how life in the physical world started on our planet Earth; this is the beginning of the era in which the whole planet crystallized in the three dimensional plane, with two divisions: White Tantra and Black Tantra.

The Atlantean civilization had seven subraces.

The first Atlantean subrace was formed by the Rmoahals. They lived in the Golden Age of Atlantis.

The second subrace was formed by the Tlavatlis, who were also living in the Golden Age and at the beginning of the Silver Age of the Atlantean epoch.

The third race of the Atlantean civilization were called the Toltecs, which still are remembered in Mexico. The Toltecs were guided by Quetzalcoatl, those great masters of the past. Toltecs means "builders," or masons.

When the fourth subrace started to appear, the Turanians, is when the black magicians started to develop, forming large groups. The Turanians were black magicians, and since nature always gives opportunities to those fallen souls in order to rise again and to develop as true human beings, the Turanians were still practicing the black magic from the Lemurian epoch, and developing witchcraft, sorcery, through Black Tantra. Those black magicians, witches, and sorcerers were developing very negative powers, controlling the animal and plant kingdoms, not only physically but internally, as they could see the elementals of nature. They were utilizing witchcraft and sorcery in order to manipulate the human masses, to conquer them. So, at that time they were worshipping the entities of darkness, demons of Klipoth, fallen angels of past cosmic days; demons such as Beelzebub were practicing Black Tantra, and were looking for proselytes. These demons were worshipped as gods, as we find in the book of Genesis:, in chapter 6, verses 1-13 that it is written:

> "And it came to pass, when Adam began to multiply on the face of the earth, and daughters were born unto them, that the Beni-Elohim saw the daughters of Adam that they were fair; and they took them wives of all which they chose (and fornicated). And Iod-Havah said, My spirit shall not always strive with Adam, for that

he also is flesh: yet his days shall be an hun-
dred and twenty years. There were giants
(the Lemuro-Atlantean Nephilim) in the
earth in those days....

"And Iod-Havah saw that the wickedness
of Adam was great in the earth, and that
every imagination of the thoughts of his
heart was only evil continually."
- Genesis 6:1-5

This is a reference to the Turanians, the
fourth subrace of the Atlantean epoch.

"And it repented Iod-Havah that he had
made Adam on the earth, and it grieved
him at his heart. And Iod-Havah said, I
will destroy Adam whom I have created
from the face of the earth; both Adam, and
beast (Behemoth), and the creeping thing,
and the fowls of the air; for it repenteth me
that I have made them." - Genesis 6:6, 7

All of this happened during the
Turanian era. At that time in the
Atlantean civilization, everywhere on that
continent was witchcraft, black magic,
and they were enjoying developing their
powers of darkness, and no one wanted
to repent, to change, and the whole Earth
was polluted. They were the outcome of

Cain, or Black Tantra, which began in the epoch of Lemuria.

So, then the Bible states:

> *"But Noah found grace in the eyes of Iod-Havah. These are the generations of Noah: Noah was a just man and perfect in his generations, and Noah walked with the Elohim."* - Genesis 6:8

Noah is a reference to what in India is called the Vaivasvata Manu, the child of the sun, the solar Manu (man, mind), a reference to those initiates from the Turanian epoch who were practicing White Tantra, as described in the Book of Enoch:

> *"After a time, my son Methuselah took a wife for his son Lamech. She became pregnant by him, and brought forth a child, the flesh of which was as white as snow and red as a rose; the hair of whose head was white like wool, and long; and whose eyes were beautiful. When he opened them, he illuminated the entire house, like the sun; the whole house abounded with light. And when he was taken from the hand of the midwife, opening also his mouth, he spoke to the Lord of righteousness. Then Lamech his father was afraid of him; and flying*

*away came to his own father Methuselah,
and said, I have begotten a son, unlike
to other children. He is not human; but,
resembling the offspring of the angels of
heaven, is of a different nature from ours,
being altogether unlike to us. His eyes are
bright as the rays of the sun; his countenance glorious, and he looks not as if he
belonged to me, but to the angels. I am
afraid, lest something miraculous should
take place on earth in his days. And now,
my father, let me entreat and request you
to go to our progenitor Enoch, and to learn
from him the truth; for his residence is with
the angels.*

"*When Methuselah heard the words of his
son, he came to me at the extremities of the
earth; for he had been informed that I was
there: and he cried out. I heard his voice,
and went to him saying, Behold, I am here,
my son; since thou art come to me. He
answered and said, On account of a great
event have I come to thee; and on account
of a sight difficult to be comprehended have
I approached thee. And now, my father,
hear me; for to my son Lamech a child has
been born, who resembles not him; and
whose nature is not like the nature of man.
His color is whiter than snow; he is redder*

than the rose; the hair of his head is whiter than white wool; his eyes are like the rays of the sun; and when he opened them he illuminated the whole house. When also he was taken from the hand of the midwife, he opened his mouth, and blessed the Lord of heaven. His father Lamech feared, and fled to me, believing not that the child belonged to him, but that he resembled the angels of heaven. And behold I am come to thee, that thou mightest point out to me the truth.

"Then I, Enoch, answered and said, The Lord will effect a new thing upon the earth. This have I explained, and seen in a vision. I have shown thee that in the generations of Jared my father, those who were from heaven disregarded the word of the Lord. Behold they committed crimes; laid aside their class, and intermingled with women. With them also they transgressed; married with them, and begot children. A great destruction therefore shall come upon all the earth; a deluge, a great destruction, shall take place in one year.

"This child who is born to you shall survive on the earth, and his three sons shall be saved with him. When all mankind who are on earth shall die, he shall be safe. And

his posterity shall beget on the earth giants, not spiritual, but carnal. Upon the earth shall a great punishment be inflicted, and it shall be washed from all corruption. Now therefore inform thy son Lamech that he who is born is his child in truth; and he shall call his name Noah, for he shall be to you a survivor. He and his children shall be saved from the corruption which shall take place in the world; from all the sin and from all the iniquity which shall be consummated on earth in his days. Afterwards shall greater impiety take place than that which had been before consummated on the earth; for I am acquainted with holy mysteries, which the Lord himself has discovered and explained to me; and which I have read in the tablets of heaven. In them I saw it written, that generation after generation shall transgress, until a righteous race shall arise; until transgression and crime perish from off the earth; until all goodness come upon it. And now, my son, go tell thy son Lamech that the child which is born is his child in truth; and that there is no deception.

"When Methuselah heard the word of his father Enoch, who had shown hint every secret thing, he returned with understand-

*ing, and called the name of that child
Noah; because he was to console the earth
on account of all its destruction."* – Book of
Enoch 105:1-20

Noah is also a reference to the atom
Nous (mind or manas). If the Elohim
were to look inside of us as we are right
now, they will find that the only good
thing that we have within us is the Atom
Nous, the Superior Manas, Manu, or
Noah. The Nous Atom, which resides
in the left ventricle of the heart, is that
spiritual atom that is always in contact
with the Christic forces, with the Monad,
the Elohim. Through that atom is how
Christ can build the real human being, if
we repent, that is, if we annihilate the ego,
that Nous Atom develops, and guides us.
This is how the Elohim guide us through
Nous (Noah, Manu) in order to rise again
as true human beings as Vaivasvatas (children of the sun).

> *"These are the generations of Noah: Noah
> was a just man (Manas, Nous) and perfect
> in his (alchemical) generations, and
> (through White Tantra) Noah walked with
> the Elohim. And Noah begat three sons,
> Shem, Ham, and Japheth."* - Genesis 6:9, 10

"Perfect in his generations" means that the way they "generated" themselves was not like that of the animals. This means that they were chaste in their sexuality, since they were not generating through fornication as animals, because they knew White Tantra. When you know White Tantra, you learn to walk with the Elohim. The Bible says that Noah walked with Elohim, meaning that through White Tantra he walked with the Elohim (the gods).

Then, the Bible states, "And Noah begat three sons, Shem, Ham, and Japheth." These are the Wedding Garment of the soul: the solar astral body, mental body, and causal body. This is how through Alchemy we have to build the three children of Noah (the Nous Atom).

Cosmically speaking, Noah represents those individuals in the Atlantean epoch who were practicing White Tantra, who knew the mysteries, the **arcana**. This is what the Bible refers to as the building of the **Ark** (Arcanum) of Noah.

> "The earth also was corrupt before the Elohim, and the earth was filled with violence. And the Elohim looked upon the earth, and, behold, it was corrupt; for all flesh

> *had corrupted his way upon the earth. And the Elohim said unto Noah, The end of all flesh is come before our face; for the earth is filled with violence before our face; and, behold, I will destroy them with the earth. Make thee an ark of gopher wood; rooms shalt thou make in the ark, and shalt pitch it within and without with pitch."* - Genesis 6:11-14

That Ark is not a ship of wood, as many people think, and who are still looking for that physical Ark. That Ark is a reference to the arcanum, the mysteries of Daath ("knowledge," of sexuality), mysteries that Noah and his followers were teaching at that time. So, they built the Ark, during which time they gathered a great army of people, and these people were prohibited from mixing with the Turanian subrace, because they were really evil, black magicians; it was forbidden. These people of the Ark-anum were the selected ones who crossed themselves with the Hyperboreans.

At that time, during the Atlantean civilization, the masters who were guiding the Atlanteans were incarnated in physical bodies, too. Those masters were the Elohim from the Hyperboreans, the second root race. Those Hyperboreans, physically speaking, came together with those people of Noah, and through White Tantra they multiplied and created the Shemites.

The Shemites were the fifth subrace of the Atlantean root race. The Shemites were prohibited to sexually cross with the Turanians that were practicing black magic, in order not to contaminate themselves with Black Tantra. This is how we have to understand the forbiddance of the Shemites to cross with the races of other Atlantean civilizations during the Atlantean epoch.

From the Shemites came Ham and Japheth, biblically speaking. Ham and Japheth represent the other subsequent subraces: Ham was the sixth, and Japheth the seventh. They were the outcome of White Tantra. Ham, the sixth subrace, is also referred to as the Arcadians, which are also mentioned in many books. Japheth, the seventh subrace, is referred to as the Mongolians, who as a subrace still exists in Asia; the present Mongolians are the outcome of that seventh subrace of the Atlantean epoch. This is why when Noah and his "family" came out of

Atlantis, they established themselves in the Gobi Desert. In that desert is precisely were our present Aryan root race began.

So, that is the story of the origins of our present Aryan root race. The Shemites, Ham, and Japheth, and all those subsequent crossings of the peoples of White Tantra, established themselves in the Gobi Desert, and there they established the basis of our present civilization, or root race, which is called the Aryan root race.

The great flood happened in order to clean the Earth of evilness, that is, the Turanians who were dedicated to fornication and black magic.

The Aryan Root Race

After the universal flood that destroyed most of the Turanians, the Golden Age of our present root race started in Tibet. At that time, the prohibition of crossing themselves with other races ended; the Shemites were then allowed to cross themselves with other Aryan races in order to give birth to the Aryan root race. So, the prohibition of the Shemites to not cross with the other races was only during the Atlantean epoch, but in the beginning of the Aryan root race, that prohibition was cancelled. This is how the Aryan root race established themselves on these continents that we know now on the planet Earth.

Many civilizations and great cities of wisdom existed in Tibet, and from Tibet, the people established in India and in China, the second subrace of our present Aryan root race. That was the Golden Age. There are many stories of past civilizations and great kings, great masters that existed in that epoch.

From China and India, the immigration of people into Chaldaea, Egypt, Babylon, Persia, Jerusalem, created the third subrace. This third subrace of the Aryan Epoch was the Silver Age, and the beginning of the Copper Age.

The Aryan civilization started to develop the Iron Age, the Kali Yuga, in the epoch of the Romans and the Greeks. The Romans and the Greeks were the outcome of the fourth subrace of this Aryan civilization. It was during that epoch that the Master Jesus of Nazareth came onto the Earth physically in order to finish his mission, to help these Monads that were entangled in too much karma.

From the Greek and Romans, the fourth subrace, the fifth subrace developed: the English, French, Germans, Polish, Russians, and all those races related with the Teutonic Anglo-Saxons in Europe, and from these peoples developed a great civilization. Still people think that the Teutonic Anglo-Saxons are the only Aryans in the world, but they are merely the fifth Aryan subrace of this great civilization.

Christopher Columbus brought European people into the lands of America; in America there were still remnants of the Atlantean civilization, and the Europeans crossed themselves with the natives of America in order to generate the sixth subrace. Some people think that the sixth subrace is about to be formed; they ignore that the sixth subrace was already formed in Latin America.

We are now at the very end of the Aryan root race. The seventh subrace is being formed right now, mainly in the United States and Canada, and how that culture is influencing the rest of the world through the mixture of all the races.

The Future Sixth Root Race

As you see, each root race ends with a great cataclysm, which "cleans the table" for the development of the next race. That is what is coming next on this planet.

At this moment, the Elohim are occupied with making seeds for the sixth root race that will come after the destruction of this Aryan root race. This current race will be destroyed just as the Turanians of Atlantis were destroyed, because of their degeneration and their practice of Black Tantra. Now though, the degeneration is worse, because, as you observe, this Aryan humanity is becoming more and more degenerated, and everybody is applauding degeneration, and welcoming degeneration.

The Elohim want to form a new civilization, a new Golden Age, but because of the present degeneration of this root race and the rottenness of the seed of this civilization, it is impossible.

When you want to grow something, you need a good seed. When you investigate the human seed of this present civilization, you see that it is no good. This is because people are dedicating themselves to fornication and sexual degeneration, through the use of drugs, not only the illicit drugs, namely, hallucinatory mushrooms, L.S.D., pills, marijuana, cocaine, alcohol, etc., that alter the sexual gene, but also the drugs, hormones, chemicals, etc. that are being proliferated through official medicine and the food supply, that are altering the glands of the physical organism, and originate a human seed that is very degenerated. Take a look around at how sickly, overweight, weak, and generally miserable humanity has become.

This is why the Elohim, the masters, seeing that most of the human seed of this Aryan root race is polluted, they are working to form the seeds for the future sixth root race through a mixture of terrestrials with extraterrestrials. Yes: there are humanities from other planets, humanities that are far superior to us in every way. Truly, this planet is a degenerated backwater, a cesspool of crime, violence, lust, greed, and all manner of filthiness. Other planets do not suffer such problems, and they want to help those of us who want to rise up out of the muck.

Still people think that there are no extraterrestrials; they believe that people do not exist on other planets. Personally, I know they exist, not because I read it, or because somebody told me, but because I have experienced it, and I do not care if people believe me or not, because to believe is not the objective of Gnosis, we have to experience it. For me, the extraterrestrials are something normal.

The extraterrestrials are interested in those people that want to change. It would be stupid if you saw a drunk or a drug addict walking down the street and you opened your doors and said "Welcome to my home." The extraterrestrials are not looking for alcoholics, drug addicts, sex addicts, etc. They are looking for people that really want to change, that really are changing psychologically; only then can you become in contact with them.

Some people believe in extraterrestrials, and some do not believe, but we Gnostics do not believe or disbelieve, we experience, and we know that one who is serious in himself, working psychologically, eventually will be in contact with the extraterrestrials, thus, it is no big deal. This is why we are always talking with certainty about these matters, because we know, and we are inviting people to know this.

This is the story of the Earth. A new root race is coming. People talk about the Golden Age, but the Golden Age that is coming is not going to be formed with prostitutes, homosexuals, lesbians, thieves, drug addicts, assassins, pedophiles, murderers, war mongers, and all the rest. The new civilization, the new root race that is coming is going to be formed with clean people, who are clean physically and psychologically. We need to be clean, because all of us are filthy; that is why we are teaching Gnosis. Thus, if we want to be chosen, we have to choose ourselves through cleaning ourselves. It is not a matter of believing, it is a matter of practicing White Tantra and meditation, in order to annihilate our filthiness.

Glossary

Aegyptopithecus: A small, tree-dwelling, fruit-eating animal living about 33MYA. Weighing about 4kg and somewhat resembling a modern-day lemur except for a full set of 32 teeth, this animal has been termed the "Dawn Ape", an important link between earlier mammals and the apes of the Miocene Epoch. Aegyptopithecus was found in the Egyptian Faiyum Depression, a rich source of Oligocene fossils.

Alcyone: 1. a third-magnitude star in the constellation Taurus: brightest star in the Pleiades. 2. Halcyon, Halcyone of Classical Mythology: a daughter of Aeolus who, with her husband, Ceyx, was transformed into a kingfisher.

Androgynous: (Greek) Gk. androgynos, "male and female in one," from andros "male" + genika "female", meaning: having male and female sexual polarities or characteristics. A sexless being capable of reproducing its own species by means of asexual reproduction, in other words, a living species that without having evident sexual organs is capable of reproducing its own species (i.e. fissiparous "Tending to break up into parts or break away from an androgynous body").

Anthropoid: (Greek: anthropoeides, "resembling a human") 1. apes 2. gorillas, orangutans, chimpanzees and mandrills resembling humans.

Australopithecus: (Literally "southern ape") A genus of extinct hominids that lived in Africa from the early Pliocene Epoch (beginning about 5.3 million years ago) to the beginning of the Pleistocene (about 1.6 million years ago). Most paleoanthropologists believe that this genus is ancestral to modern human beings. The australopithecines were presumably distinguished from early apes by their upright posture and bipedal gait. Their brains were relatively small, not very different from those of living apes, but their teeth were more similar to those of humans.

Brachycephalus: (Greek) short-headed or broad-headed.

Briareos: (Greek) A monstrous creature with one-hundred arms, the offspring of Gaia (earth) and Uranus (sky). Sometimes Poseidon is mentioned as his father.

Catarrhine: A term describing those belonging or pertaining to the group Catarrhini, comprising humans, anthropoid apes, and Old World monkeys, having the nostrils close together and opening downward and a nonprehensile, often greatly reduced or vestigial tail.

Chaos: (Greek) The abyss (not the inferior abyss), or the "great deep." Personified as the Egyptian Goddess Neith. The Great Mother, the Immaculate Virgin from which arises all matter. The Chaos is WITHIN the Ain Soph.

Consciousness: "Wherever there is life, there exists the consciousness. Consciousness is inherent to life as humidity is inherent to water." —Samael Aun Weor, *Fundamental Notions of Endocrinology and Criminology*

From various dictionaries: 1. The state of being conscious; knowledge of one's own existence, condition, sensations, mental operations, acts, etc. 2. Immediate knowledge or perception of the presence of any object, state, or sensation. 3. An alert cognitive state in which you are aware of yourself and your situation. In Universal Gnosticism, the range of potential consciousness is allegorized in the Ladder of Jacob, upon which the angels ascend and descend. Thus there are higher and lower levels of consciousness, from the level of demons at the bottom, to highly realized angels in the heights.

"It is vital to understand and develop the conviction that consciousness has the potential to increase to an infinite degree." —The 14th Dalai Lama

"Light and consciousness are two phenomena of the same thing; to a lesser degree of consciousness, corresponds a lesser degree of light; to a greater degree of consciousness, a greater degree of light." —Samael Aun Weor, *The Esoteric Treatise of Hermetic Astrology*

Demiurge: (Greek, for "worker" or "craftsman") The Demiurgos or Artificer; the supernal power that built the universe. Freemasons derive from this word their phrase "Supreme Architect." Also the name given by Plato in a passage in the *Timaeus* to the creator God.

Dhyan Chohan: (Sanskrit) "Lord of the Light." A Cosmocreator or Elohim. The Divine Intelligences supervising the cosmos. "A Dhyan Chohan is one who has already abandoned the four bodies of sin, which are the physical, astral, mental and causal bodies. A Dhyan Chohan only acts with his Diamond Soul. He has already liberated himself from Maya (illusion); thus, he lives happily in Nirvana." - Samael Aun Weor, *The Revolution of Beelzebub*

Elliot, W. Scott: author of "The Story of Atlantis," 1896.

Elohim: [אלהים] An Hebrew term with a wide variety of meanings. In Christian translations of scripture, it is one of many words translated to the generic word "God," but whose actual meaning depends upon the context. For example:

1. In Kabbalah, אלהים is a name of God the relates to many levels of the Tree of Life. In the world of Atziluth, the word is related to divnities of the sephiroth Binah (Jehovah Elohim, mentioned especially in Genesis), Geburah, and Hod. In the world of Briah, it is related beings of Netzach and Hod.

2. El [אל] is "god," Eloah [אלה] is "goddess," therefore the plural Elohim refers to "gods and goddesses," and is commonly used to refer to Cosmocreators or Dhyan-Choans.

3. אלה Elah or Eloah is "goddess." Yam [ים] is "sea" or "ocean." Therefore אלהים Elohim can be אלה-ים "the sea goddess" [i.e. Aphrodite, Stella Maris, etc.]

There are many more meanings of "Elohim." In general, Elohim refers to high aspects of divinity.

"Each one of us has his own Interior Elohim. The Interior Elohim is the Being of our Being.

The Interior Elohim is our Father-Mother. The Interior Elohim is the ray that emanates from Aelohim." - Samael Aun Weor, *The Gnostic Bible: The Pistis Sophia Unveiled*

Euclid: A Greek mathematician circa 300 B.C. who wrote about three-dimensional geometry.

Evolution: A concept that embodies the belief that existing animals and plants developed by a process of gradual, continuous change from previously existing forms. This theory, also known as "descent with modification," constitutes organic evolution. Inorganic evolution, on the other hand, is concerned with the development of the physical universe from unorganized matter. Organic evolution, as opposed to belief in the special creation of each individual species as an immutable form, conceives of life as having had its beginnings in a simple primordial protoplasmic mass (probably originating in the sea) from which, through the long eras of time, arose all subsequent living forms.

"The scientists of our world think that evolution is gradual, but this is not so, for there are forced movements in nature, both backwards and forwards, similar to the tides of the sea. The movements which built up Greece, and the knowledge given to Egypt of the forces of nature, were all measured. The Renaissance, and the retrograde movements which came after, were measured, and even now we can see the path of a new civilization, and measure its dark age of destruction and turmoil, and also its emergence into peace and security in the future." —M, The Lord God of Truth Within

Gemmation: The formation of a new individual, either animal or vegetable, by a process of budding; an asexual method of reproduction; gemmulation; gemmiparity.

Gnosis: (Greek) Knowledge.

1. The word Gnosis refers to the knowledge we acquire through our own experience, as opposed to knowledge that we are told or believe in. Gnosis - by whatever name in history or culture - is conscious, experiential knowledge, not merely intellectual or conceptual knowledge, belief, or theory. This term is synonymous with the Hebrew "daath" and the Sanskrit "jna."

2. The tradition that embodies the core wisdom or knowledge of humanity.

"Gnosis is the flame from which all religions sprouted, because in its depth Gnosis is religion. The word "religion" comes from the Latin word "religare," which implies "to link the Soul to God"; so Gnosis is the very pure flame from where all religions sprout, because Gnosis is Knowledge, Gnosis is Wisdom." —Samael Aun Weor, *The Esoteric Path*

"The secret science of the Sufis and of the Whirling Dervishes is within Gnosis. The secret doctrine of Buddhism and of Taoism is within Gnosis. The sacred magic of the Nordics is within Gnosis. The wisdom of Hermes, Buddha, Confucius, Mohammed and Quetzalcoatl, etc., etc., is within Gnosis. Gnosis is the Doctrine of Christ." —Samael Aun Weor, *The Revolution of Beelzebub*

Hermaphrodite: (Greek) In the pure esoteric tradition (not the modern, degenerated remnants), hermaphrodite means "a son of Hermes and Aphrodite" (Herm-Aphrodite), one who physically develops the brain power, objective reasoning (Hermes) by means of the transmutation of its own sexual libido (Aphrodite).

Physically, hermaphrodite refers to the human being of the Lemurian epoch, who developed both male and female sexual organs and characteristics in the physical body and who was capable of self proliferating its species without the necessity of sexual intercourse. Thus, a true hermaphrodite is a human being who physically produces sperm and ovum within their masculine-feminine sexual genitalia. In order to create, they fecundate themselves (in a manner similar to other hermaphrodite creatures); they physically unite the outcome of their own two sexual polarities (sperm and ovum) by means of willpower.

Human Being: In general, there are three types of human beings:

1. The ordinary person (called human being out of respect), more accurately called the intellectual animal

2. The true human being or man (from manas, mind): someone who has created the soul (the solar bodies), symbolized as the chariot of Ezekiel or Krishna, the Wedding Garment of Jesus, the sacred weapons of the heroes of mythology, etc. Such persons are saints, masters, or buddhas of various levels.

3. The Superman: a true human being who has also incarnated the Cosmic Christ, thus going beyond mere sainthood or buddhahood, and into the highest reaches of liberation. These are the founders of religions, the destroyers of dogmas and traditions, the great rebels of spiritual light.

According to Gnostic anthropology, a true human being is an individual who has conquered the animal nature within and has thus created the Soul, the Mercabah of the Kabbalists, the Sahu of the Egyptians, the To Soma Heliakon of the Greeks: this is "the body of gold of the solar man." A true human being is one with the Monad, the Inner Spirit. It can be said that the true human being or man is the inner Spirit (in Kabbalah, Chesed. In Hinduism, Atman).

"Every spirit is called man, which means that only the aspect of the light of the spirit that is enclothed within the body is called man. So the body of the spirit of the holy side is only a covering; in other words, the spirit is the actual essence of man and the body is only its covering. But on the other side, the opposite applies. This is why it is written: "you have clothed me with skin and flesh..." (Iyov 10:11). The flesh of man is only a garment covering the essence of man, which is the spirit. Everywhere it is written the flesh of man, it hints that the essence of man is inside. The flesh is only a vestment for man, a body for him, but the essence of man is the aspect of his spirit." —Zohar 1A:10:120

Intellectual Animal: The current state of humanity: animals with intellect.

When the Intelligent Principle, the Monad,
sends its spark of consciousness into Nature,
that spark, the anima, enters into manifestation
as a simple mineral. Gradually, over millions of
years, the anima gathers experience and evolves
up the chain of life until it perfects itself in the
level of the mineral kingdom. It then graduates
into the plant kingdom, and subsequently into
the animal kingdom. With each ascension the
spark receives new capacities and higher grades
of complexity. In the animal kingdom it learns
procreation by ejaculation. When that animal
intelligence enters into the human kingdom, it
receives a new capacity: reasoning, the intellect;
it is now an anima with intellect: an Intellectual
Animal. That spark must then perfect itself in
the human kingdom in order to become a complete and perfect human being, an entity that
has conquered and transcended everything that
belongs to the lower kingdoms. Unfortunately,
very few intellectual animals perfect themselves;
most remain enslaved by their animal nature,
and thus are reabsorbed by Nature, a process
belonging to the devolving side of life and
called by all the great religions "Hell" or the
Second Death.

"The present manlike being is not yet human;
he is merely an intellectual animal. It is a very
grave error to call the legion of the "I" the
"soul." In fact, what the manlike being has is
the psychic material, the material for the soul
within his Essence, but indeed, he does not have

a Soul yet." —Samael Aun Weor, *The Revolution of the Dialectic*

"Whosoever possesses the physical, astral, mental, and causal bodies receives the animistic and spiritual principles, and becomes a true human being. Before having them, one is an intellectual animal falsely called a human being. Regarding the face and shape of the physical body of any intellectual animal, they look like the physical characteristics of a human being, however, if their psychological processes are observed and compared with the psychological processes of a human being, then we find that they are completely different, totally distinct." —Samael Aun Weor, *Kabbalah of the Mayan Mysteries*

"I died as a mineral and became a plant,
I died as plant and rose to animal,
I died as animal and I was Man.
Why should I fear? When was I less by dying?
Yet once more I shall die as Man, to soar
With angels blest; but even from angelhood
I must pass on: all except God doth perish.
When I have sacrificed my angel-soul,
I shall become what no mind e'er conceived.
Oh, let me not exist! for Non-existence
Proclaims in organ tones, To Him we shall
return." —Jalal al-Din Muhammad Rumi (1207 – 1273) founder of the Mevlevi order of Sufism

Jiva: (Sanskrit) Sanskrit, literally "life, living being, the principle of life, vital breath, soul, existing, existence." In Hinduism, jiva is often a reference only to the individual, embodied soul.

Kumaras: (Sanskrit) From H.P. Blavatsky: "The first Kumaras are the seven sons of Brahma.

It is stated that their name was given to them owing to their formal refusal to 'procreate their species,' and so they 'remained Yogis,' as the legend says."

Lacertid: Any of numerous Old World lizards of the family Lacertidae.

Law of the Eternal Heptaparaparshinokh: The Law of Seven. This is the fundamental principle of the organization of everything that is created. Thus we have seven primary colors, seven primary chakras, seven primary notes, seven primary planets, etc.

Limbus: "Limbus Major" is a term used by Paracelsus; it means "primordial matter."

Logos: (Greek, "word") In Greek and Hebrew metaphysics, the unifying principle of the world. The central idea of the Logos is that it links God and man, hence any system in which the Logos plays a part is monistic. The Logos is the manifested deity of every nation and people; the outward expression or the effect of the cause which is ever concealed. Thus, speech is the Logos of thought; hence it is aptly translated as the Verb, the Word.

Manu: (Sanskrit) From Hindu mythology, the progenitor and lawgiver of the human race. From H.P. Blavatsky: "Who was Manu, the son of Swayambhuva? The secret doctrine tells us that this Manu was no man, but the representation of the first human races evolved with the help of the Dhyan-Chohans (Devas, Elohim) at the beginning of the first round. But we are told in his Laws (Book I. 80) that there are fourteen

Manus for every Kalpa -- or interval from
creation to creation -- and that in the present
divine age, there have been as yet seven Manus...
We are told in the Sacred Hindu scriptures that
the first Manu produced six other Manus (seven
primary Manus in all), and these produced in
their turn each seven other Manus (Bhrigu I,
61-63) -- the production of the latter standing
in the occult treatises as 7 x 7. Just as each
planetary Round commences with the appearance of a 'Root Manu' (Dhyan Chohan) and
closes with a 'Seed-Manu,' so a Root and a Seed
Manu appear respectively at the beginning and
the termination of the human period on any
particular planet."

Materialist: In the esoteric sciences, a "materialist"
is one who only believes in what his five senses
can tell him, thus he relies exclusively on the
data of the sensory, third dimensional world.
This type of person has no understanding of
the superior senses or the superior dimensions,
and thus is limited to what he can perceive
physically. As C.W. Leadbeter said, "It is one of
the commonest of mistakes to consider that the
limit of our power of perception is also the limit
of all there is to see."

Miocene epoch: The fourth epoch of the Tertiary
period in the Cenozoic era of geologic time (see
Geologic Timescale, page 120), lasting from
around 24.6 to 5.1 million years ago.

Monera: The taxonomic kingdom that comprises
the prokaryotes (bacteria and cyanobacteria).
Prokaryotes are single-celled organisms that
lack a membrane-bound nucleus and usually

lack membrane-bound organelles (mitochondria, chloroplasts). They have a small ring of DNA as their genetic material and reproduce asexually. In the theories of Haeckel, all life presumably arose from a singular root Monera, afloat in a primordial sea.

Natural Selection: The process by which forms of life having traits that better enable them to adapt to specific environmental pressures, as predators, changes in climate, or competition for food or mates, will tend to survive and reproduce in greater numbers than others of their kind, thus ensuring the perpetuation of those favorable traits in succeeding generations. Cf. "survival of the fittest."

Ontogeny: The development or developmental history of an individual organism. Ernst von Haeckel proposed that "ontogeny recapitulates phylogeny." This jargon, when translated into English, asserts that as an embryo develops it passes through stages that are equivalent to the adult forms of its ancestors. For example, according to Haeckel, a human embryo would pass through a stage in which it has features of an adult fish, then features of an adult amphibian, and so forth. Gnostic anthropology observes the three lower kingdoms in the development of the human fetus, a recapitulation of the millions of years of the evolution of the consciousness as it ascended up the evolutionary path from mineral to plant to animal, until finally achieving its goal: the opportunity of a human birth.

Ophidioid: Of or pertaining to the Ophidiidæ, a family of fishes which includes many slender species.

Paleolitical: the period B.C. 20,000-17,000. See chart on page 120.

Pelasgians: A prehistoric people inhabiting Greece, Asia Minor, and the islands of the eastern Mediterranean. Mentioned by Heroditus in his Histories.

Phylogeny: 1. the development or evolution of a particular group of organisms. 2. the evolutionary history of a group of organisms, esp. as depicted in a family tree.

Pithecus-Noah: Pithecus is from australopithecus ("southern ape") which is the materialistic anthropologists name for an early hominid. Noah is a Biblical and mythological character, the primary human survivor of the world flood. According to the Bible, Noah had three sons; this is interpreted literally by the common religious view and is seen as the origin of all contemporary humanity. However, some propose the theory that three early hominid skeletons (australopithecus africanus, australopithecus robustus, australopithecus boisei) are our ancestors, and this supposedly ties in with the three sons of Noah.

Pleiades: 1. Classical Mythology: seven daughters of Atlas and half sisters of the Hyades, placed among the stars to save them from the pursuit of Orion. One of them (the Lost Pleiad) hides, either from grief or shame. 2. Astronomy: a conspicuous group or cluster of stars (M45) in the constellation Taurus, commonly spoken of as seven, though only six are visible. The cluster consists of some 500 stars, has a diameter of 35 light-years, and is 400 light-years distant from the earth. Six stars are easily visible to the naked eye—Alcyone (the brightest), Electra, Celaeno, Sterope, Maia, and Taygete. Known as the Seven Sisters, this group was named by the Greeks for the seven daughters of Atlas and Pleione; the seventh Pleiad was, according to legend, lost or in hiding. Many faint stars associated with the other six are visible with the telescope; one of these stars may have been much brighter and visible to the naked eye in ancient times, thus accounting for the many early references to seven stars. The Pleiades cluster is 150 million years old, making it a young star cluster.

Prajnaparamita: (Sanskrit) Literally, "the perfection of wisdom." The Prajnaparamita type of vision in its more elevated degree is the same "Eye of Dangma," the Polyvision that one acquires when the ego is 100% dead. The Prajnaparamita Sutra, also known as the Heart Sutra, is the source of the famous mantra Gate Gate Paragate Parasamgate Bodhi Swaha.

Protist: Any of various one-celled organisms, classified in the kingdom Protista, that are either free-living or aggregated into simple colonies and that have diverse reproductive and nutritional modes, including the protozoans, eukaryotic algae, and slime molds. Some classification schemes also include the fungi and the more primitive bacteria and blue-green algae or may distribute the organisms between the kingdoms

Plantae and Animalia according to dominant characteristics.

Psychological Space: The different dimensions through which the Psyche or consciousness (the essence-part of the Monad) has developed itself through different evolving or devolving organisms within the ethereal, astral, mental, causal, or divine dimensions. For Psyche we understand the element which allow us to be cognizant of the dimension where it is active through physical, ethereal, astral, mental, causal, or divine bodies.

Seiche: (pronounced Sigh-shh) The word was introduced to science by the Swiss seismologist F.A. Forel in 1890. The word had apparently long been used in the German-speaking part of Switzerland to describe oscillations in alpine lakes.

Sirius: The Dog Star, the brightest star in the sky. It is located in the constellation Canis Major; its Bayer designation is Alpha Canis Majoris. Sirius [Greek, "scorching") is exceeded in brightness only by the sun, the moon, and Venus and by Mars and Jupiter at their maximum brightness. Sirius is about twice the size of the sun and about 20 times as luminous. It is also one of the nearest stars, lying at a distance of 8.7 light-years, so that it has been studied extensively. From an analysis of its motions, F. W. Bessel concluded (1844) that it had an unseen companion, which was later (1862) confirmed by observation. The companion, Sirius B, is a white-dwarf star and has also been the object of considerable study because it is the first white dwarf whose spectrum was found to exhibit a gravitational red shift as predicted by the general theory of relativity .

Three-brained biped: Gnostic psychology recognizes that humanoids actually have three centers of intelligence within: an intellectual brain, an emotional brain, and a motor/instinctive/sexual brain. These are not physical brains; they are divisions of organized activity. Each one functions and operates independent of the others, and each one has a host of jobs and duties that only it can accomplish. Of course, in modern humanity the three brains are grossly out of balance and used incorrectly. See Revolutionary Psychology by the author.

Turanian People: a group of Atlanteans that survived the Deluge.

Index

About the Author

His name is Hebrew סמאל און ואור, and is pronounced "sam-ayel on vay-or." You may not have heard of him, but Samael Aun Weor changed the world.

In 1950, in his first two books, he became the first person to reveal the esoteric secret hidden in all the world's great religions, and for that, accused of "healing the ill," he was put in prison. Nevertheless, he did not stop. Between 1950 and 1977 – merely twenty-seven years – not only did Samael Aun Weor write over sixty books on the most difficult subjects in the world, such as consciousness, kabbalah, physics, tantra, meditation, etc., in which he deftly exposed the singular root of all knowledge — which he called Gnosis — he simultaneously inspired millions of people across the entire span of Latin America: stretching across twenty countries and an area of more than 21,000,000 kilometers, founding schools everywhere, even in places without electricity or post offices.

During those twenty-seven years, he experienced all the extremes that humanity could give him, from adoration to death threats, and in spite of the enormous popularity of his books and lectures, he renounced an income, refused recognitions, walked away from accolades, and consistently turned away those who would worship him. He held as friends both presidents and peasants, and yet remained a mystery to all.

When one reflects on the effort and will it requires to perform even day to day tasks, it is astonishing to consider the herculean efforts required to accomplish what he did in such a short time. But, there is a reason: he was a man who knew who he was, and what he had to do. A true example of compassion and selfless service, Samael Aun Weor dedicated the whole of his life to freely helping anyone and everyone find the path out of suffering. His mission was to show all of humanity the universal source of all spiritual traditions, which he did not only through his writings and lectures, but also through his actions.

Your book reviews matter.

Glorian Publishing is a very small non-profit organization, thus we have no money to spend on marketing and advertising. Fortunately, there is a proven way to gain the attention of readers: book reviews. Mainstream book reviewers won't review these books, but you can.

The path of liberation requires the daily balance of three active factors:

- · birth of virtue
- · death of vice
- · sacrifice for others

Writing book reviews is a powerful way to sacrifice for others. By writing book reviews on popular websites, you help to make the books more visible to humanity, and you might help save a soul from suffering. Will you do your part to help us show these wonderful teachings to others? Take a moment today to write a review.

Donate

Glorian Publishing is a non-profit publisher dedicated to spreading the sacred universal doctrine to suffering humanity. All of our works are made possible by the kindness and generosity of sponsors. If you would like to make a tax-deductible donation, you may send it to the address below, or visit our website for other alternatives. If you would like to sponsor the publication of a book, please contact us at (844) 945-6742 or help@gnosticteachings.org.

Glorian Publishing
PO Box 110225
Brooklyn, NY 11211 US
Phone: (844) 945-6742

VISIT US ONLINE AT gnosticteachings.org